Zoophysiology and Ecology
Volume 6

Klaus Schmidt-Koenig

Migration and Homing in Animals

With 64 Figures

Springer-Verlag
Berlin Heidelberg New York 1975

Prof. Dr. KLAUS SCHMIDT-KOENIG

Universität Tübingen, Institut für Biologie III,
Lehrstuhl Zoophysiologie, Abteilung Verhaltensphysiologie
74 Tübingen, Beim Kupferhammer 8

4/1976
Biol. Cent.

Science Center
QL
754
S35

Cover Motif: Animal migration is one of the most fascinating environmental phenomena. In addition to the well known birds very many other animals also cover respectable distances during migration.

ISBN 3-540-07433-3 Springer-Verlag Berlin Heidelberg New York
ISBN 0-387-07433-3 Springer-Verlag New York Heidelberg Berlin

Library of Congress Cataloging in Publication Data. Schmidt-Koenig, K., 1930–. Migration and homing in animals. (Zoophysiology and Ecology; v. 6). Bibliography: p. Includes index. 1. Animal migration. 2. Animal orientation. I. Title. QL754.S35.591.5′2.75–25594.

Typesetting and printing: Zechnersche Buchdruckerei, Speyer.
Bookbinding: Brühlsche Universitätsdruckerei, Gießen.

Preface

On being asked to write a book on migration and homing in animals, intended as an introductory text to inform and stimulate both students and non-specialists, I saw the following alternatives for an outline:

(1) I could discuss known or hypothetical mechanisms of orientation, and enumerate animals known or thought to use these mechanisms in migration and homing.

(2) I could discuss the known feats of orientation by animals under field observation (e. g. migration), following some practicable order, with a subsequent discussion of the attempts at and possible success in elucidating the basic mechanisms of orientation.

Both alternatives have obvious disadvantages. (1) would suffer from the fact (a) that very few orientation mechanisms (such as the sun, star, or magnetic compass) have been firmly established and (b) that for many animals the modes of orientation are unknown; therefore, for many animals whose considerable feats of orientation are well known, an appropriate allocation could not be made. With (2) the disadvantage is that due to the complexity of animal movements, it is difficult to find some relevant order and that in many animal groups discussions of certain known mechanisms such as the sun compass would recur. I have selected outline (2) as the much lesser evil. The discussion pursues a compromise between taxonomic order and similarity in feats of orientation or methodological approaches chosen by the various investigators. I shall by no means attempt to cover all animals or mention all authors in the field of orientation, but rather shall select examples from major groups of animals of particular interest, which have attracted attention in the short history of orientation research.

The discussion follows a simple scheme for each major group: A. Field performance in orientation (descriptive section), B. Experimental and theoretical analysis of the underlying mechanisms (analytical section). A strict division between A. and B. is, however, not always possible, descriptive and analytical aspects overlapping to some degree. There are groups of animals in which much more has been reported than can be discussed here on both field performance and on experimental and theoretical analysis, for example in fishes and in birds. In other groups little is reported in one or even in both categories. Length and depth of coverage will, therefore, necessarily be inhomogenous, resulting in correspondingly inhomogenous chapters. Furthermore, whereas in some fields modern instrumentation is employed and is here mentioned, in other fields classical techniques still provide the major source of the evidence available.

Many animals perform long-distance migration, the best known being whales, birds, turtles, fishes and butterflies, that migrate over hundreds, thousands or even tens of thousands of kilometers. However, there are numerous other animals that also move over distances considerable even by human measurement, and enormous for the means of locomotion of the migrants. Even if the movements of animals such as spiders, toads, bees, sand hoppers, and the like are not so long and impressive by human standards, they may nevertheless require accurate orientation abilities. Many animal movements are on a relatively small scale such as those described for example by FUNKE (1968)[1] in *Patella* or by SIEBECK (1960; 1968; 1973) in freshwater crustaceans, both of which return to previously inhabited locations. These and many others, although qualifying as homing, will not be discussed in detail here, the emphasis being more on extensive animal movements.

Homing abilities enable animals to reach and/or to return to some specified location in a non-random, i.e., directed fashion. The goal of an outward journey or the location for return need not necessarily be a geographically fixed spot such as a specific breeding pond for toads, or a specific hive for bees, it may also be a habitat such as a certain part of the shore suitable for a spider to live, a forest for a cock chafer to mate, or a northern or southern area where milkweed provides the exclusive food for the monarch butterfly caterpillars. While some forms of migration result from passive drifts by currents or wind (eel larvae or locusts), many forms of migration and homing certainly require true navigational abilities. True navigation may be defined as the ability to find a goal when released in unfamiliar surroundings without utilizing known landmarks (SCHMIDT-KOENIG, 1965; 1970a). This is neither the place to discuss necessary subdivisions or possible other definitions of navigation nor to argue over the implications and weaknesses of this definition, originally evolved for birds (GRIFFIN, 1952), nor to decide whether this definition may hold true for all other animals showing some considerable

---

[1] FUNKE (1968) also presented a table of work on homing in other gastropods.

movement, "considerable movement", in turn, requiring definition. One major reason for not discussing definitions here is that extremely few systems of goal-finding have, to date, been firmly established (e. g. in bees); most are as yet entirely obscure. Statements as to whether some examples of well-known orientational accomplishment would qualify as navigation in some specific definition would, therefore, be premature. I prefer to utilize the term navigation in a way similar to that in which the terms "hormone" or "species" are used: although a strict definition is extremely difficult or even impossible, one can work with the terms "hormone", or "species" quite satisfactorily.

Tübingen,
December 1975   K. SCHMIDT-KOENIG

Acknowledgements

Large parts of this book were written while the author was supported by grants from the National Aeronautic and Space Administration (Grant NGR 34-001-019) in the Department of Zoology, Duke University, Durham, N. C. USA, where the easily accessible and well-stocked library was of considerable help. While at the University of Göttingen, the author held research grants from the Deutsche Forschungsgemeinschaft whose support is also gratefully acknowledged. I am indebted to the editors of this series, D. S. FARNER (Seattle) and H. LANGER (Bochum) and to J. KIEPENHEUER (Tübingen) for their helpful comments on the manuscript and to E. BATSCHELET (Zürich) for his critical comments on the Appendix: "Some Statistical Methods for the Analysis of Orientation Data". I am grateful for the many helpful discussions with a number of friends and colleagues too large to be enumerated, and thank also my assistant HANNELORE TABEL for her untiring help in typing the manuscripts and for skilfully preparing the illustrations.

Contents

1. Crustaceans and Spiders

A. Field Performance in Orientation

Little is known about migratory movements of crustaceans. Probably typical for many species are the movements of the Florida spiny lobster *(Panulirus argus)* described by DAWSON and IDYLL (1951). The authors interpret their observations on tagged animals as indicative of random wandering since 90 % of a total of 251 recoveries were within not more than 30 km from the site

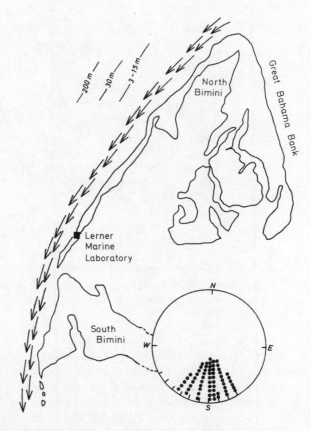

Fig. 1. Migration of spiny lobsters *(Panulirus argus)* near Bimini, Bahamas, observed during 5 days in November, 1969. The inserted diagram gives the bearings of 1,260 animals; each dot symbolizes 20 lobsters. From HERRNKIND (1970)

of release; only few individual specimens were recovered up to 200 km away. There was no evidence of seasonal movements along the Florida coast. In contrast, HERRNKIND (1969, 1970) reported rather massive fall migratory movements of spiny lobsters off Bimini in the Bahamas, the lobsters heading south (Fig. 1) in queues typical for this species when migrating. Neither origin nor destination were known. Similarly, coastal populations of the northern lobster (*Homarus americanus*) are, in general, non-migratory, while populations inhabiting the outer continental shelf migrate seasonally (COOPER and UZMANN, 1971). Experiments on tagged and displaced animals indicated well-oriented shoalward migration in spring and return to the edge of the continental shelf in fall or winter (Fig. 2). All authors speculate on the mode of orientation but conclusive experiments have not yet been made.

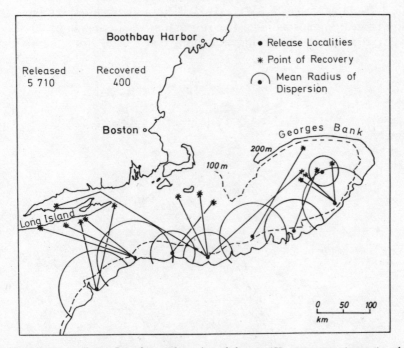

Fig. 2. Mean dispersion of tagged northern lobsters (*Homarus americanus*) and some examples of the longest shoalward migration. Of 5,710 tagged animals 400 were recovered. After COOPER and UZMANN (1971)

Several taxa of terrestrial crabs in tropical regions of America, Africa and the Indo-Pacific area migrate seasonally to water in order to breed. Detailed and quantitative accounts of these movements are not available to my knowledge. Better known and suitable for discussion here are "short-distance" movements of littoral amphipods for example. The sand hopper (*Talitrus saltator*) lives on European shores in wet sand just above the high-water mark. According to PAPI (1960) the animals emerge at night from their subterranean refuges and move several tens, occasionally up to 100 m inland. These movements occur

2

irregularly depending upon certain favorable environmental and meteorological conditions such as high atmospheric humidity and medium ambient temperature. The number of migrating animals may be very large: 11,429 specimens were trapped within 37 days during September and October in a V-shaped trap with a seaward opening of 185 cm placed about 25 to 30 m from the normal habitat. It is not known whether the inland migration is a directed or a more or less random movement. The return seaward, however, is directed; the animals move along the direct route perpendicular to the shoreline. This seaward-directed behavior was and is the methodological basis for a large number of experiments aimed at explaining the orientation mechanisms used by the animals.

B. Experimental and Theoretical Analysis

If *Talitrus* or related forms (*e. g., Talorchestia deshayesei, Orchestia mediterranea; Orchestoidea corniculata*) are displaced to an environment drier than their normal habitat, then they try to move seaward; if displaced to an environment wetter than their normal habitat, then they try to move towards the land. Such oriented movements may justifiably be classified as "homing" to some specified home territory or ecological area. Without actual migration being known or being likely, a number of other littoral or inner-bank-inhabiting arthropods offer very similar methodologically useful behavior patterns; mainly the isopods, *Tylos latreilii* and *Idotea baltica;* the carabid, *Omophron limbatum;* the staphylinid, *Paederus rubrothoracicus;* the tenebrionid, *Phaleria provencialis;* spiders of the genus *Arctosa*, and a few others. They are, therefore, mentioned here. Other beetles, such as for example the cockchafer *Melolontha*, show different orientation reactions which have been investigated in a different manner and are, therefore, not included here.

The pioneers in this field, L. PARDI and F. PAPI, developed two slightly different methods for investigating the orientation behavior of *Talitrus saltator*. Both methods have in common that 25–30 specimens are placed together in a dry dish, thus triggering seaward flight. The method of recording differed, as the animals were (a) either man-released in the center and then individually trapped on adhesive tape at the periphery of an arena 3 m in diameter, or (b) were kept in a jar, where their distribution was photographed periodically (Figs. 3 and 4). By present standards, release of single specimens, each specimen only once, and trapping at the periphery would be appropriate for subsequent statistical analysis, e. g. the calculation of a mean vector (e. g., by vector analysis, an approach also pioneered by PARDI and PAPI). However, these authors supplied such a wealth of data, and their results were in most cases so consistent that their general conclusions are widely accepted (exception see p. 12). The basic observation was: independent of distance and direction of displacement, each population headed in that compass direction that would have been the direct route to the beach at home (Fig. 5).

3

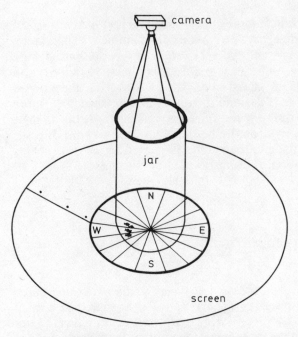

Fig. 3. Experimental arrangement as developed by PARDI and PAPI for photographic recording of escape attempts of *Talitrus* kept in a jar upside down. A camera is mounted above the jar. After PARDI and PAPI (1953)

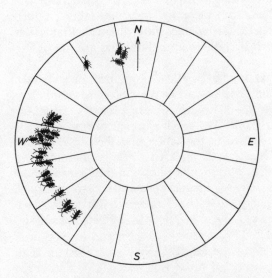

Fig. 4. Reproduction of an experimental distribution (seen from above) of *Talitri* as observed and photographed by PARDI and PAPI. The animals expected the sea to the west. From PARDI and PAPI (1953)

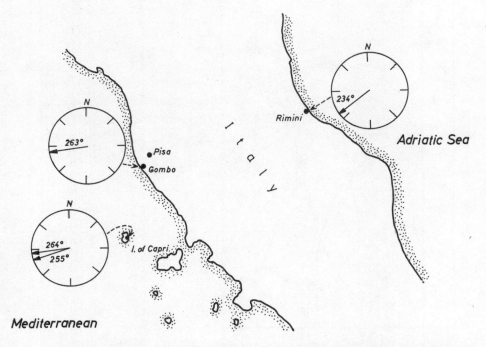

Fig. 5. Mean directions (arrows in the circular diagrams) of *Talitri* taken from Gombo beach (running N–S facing west) to beaches facing east at Rimini and the Isle of Capri. From PARDI and PAPI (1953)

When displaced from their home area on beaches facing west to beaches facing east, the animals, nevertheless, tried to escape to the west, i.e., away from the life-saving water. Thus, the direction is chosen independently of the actual position of the sea. Different populations with differently oriented shorelines have correspondingly different escape directions (Fig. 6). The orientation of members of populations living for example, at the tip of a small peninsula with a beach whose arc may be more than 180° is, however, not entirely clear.

The sun was found to be the guiding factor in the following manner: when sand hoppers in the experimental arena or jar were screened from a direct view of the sun but the sun was reflected by a mirror from the opposite direction (a method first introduced by SANTSCHI, 1911 with ants), the animals also turned 180° (PARDI and PAPI, 1952). Furthermore, when exposed to an artificial light (flashlight) in a closed room, the animals tried to escape approximately in the angle from the flashlight that would have corresponded outside with the angle of the sun. Finally, under overcast skies and also on moonless nights, no directed behavior could be observed.

The animals compensate for the (apparent) movement of the sun in the sky with the help of a time sense, the so-called internal clock, and a "knowledge" of the sun's movement. These provide the basis of the so-called sun compass, an ability that has been demonstrated in many species of animals. The interaction of the internal clock and sun movement has been demonstrated in several experi-

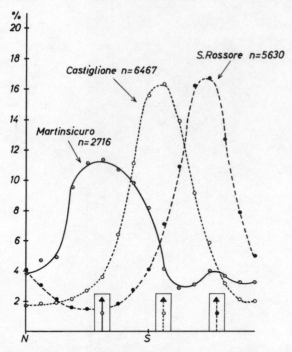

Fig. 6. Frequency distributions (ordinate in %) of three different populations of *Talitrus saltator* with differently oriented shore lines. The theoretical escape directions (perpendicular to the respective shore line) are indicated as arrows on the abscissa (compass direction), *n* indicates the total number of counts. After PARDI (1960)

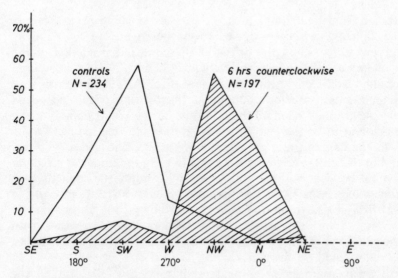

Fig. 7. Frequency distribution (ordinate in %) of *Talorchestia deshayesei* from Laragua in true compass directions (abscissa). The theoretical flight direction of controls is 220°, the theoretical direction of experimentals is 220° + 90° = 310°. The mean directions actually observed are 227° and 327°, respectively, largely in agreement with the theory. After PARDI (1957)

ments. PARDI (1957) shifted the internal clock of groups of *Talitrus* or *Talorchestia* by exposing them, in a light-proof chamber, to an artificial light-dark cycle reset from local time by 6 h counter-clockwise, or by 12 h. As may be seen from Fig. 7, a 6 h shift counter-clockwise of the internal clock resulted in a deviation in direction by the experimental animals of roughly 90° clockwise from that of control animals. The reason is that for the experimental animals in a light-dark cycle shifted counter-clockwise by 6 h, it is for example, 12 noon, when it is actually 12 + 6 h = 6 p.m. local solar time (Fig. 8). At 12 noon, the experimental

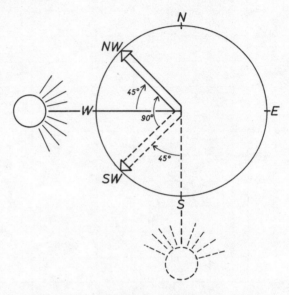

Fig. 8. Theoretical diagram to illustrate the result of a clock shift of 6-h counter-clockwise on the sun-compass orientation of an animal normally heading SW

animal would expect the sun to be due south. In order to pursue its usual direction of movement (southwest) it would be necessary to bear 45° to the right of the sun. In reality at this time (6 p.m. local time), the sun is actually roughly west, and the animal, bearing 45° to the right of it, is heading northwest, 90° clockwise of the usual flight direction. Corresponding results were obtained when the animals were kept at low temperatures of 0° to +3 °C for a specified number of hours rather than being exposed to a shifted artificial day-night regime, showing that low temperatures may stop the clock. Furthermore, PAPI (1955a) was able to demonstrate that one group of *Talitrus*, when tested immediately after large geographical displacement from Italy to Argentina, still oriented according to the sun as if it were the sun at home and did not adjust to the local conditions in South America. The sun in Argentina moved counter-clockwise, a feature ignored by the animals which obviously reacted only to its actual position. This type of experiment may be considered as a classical approach demonstrating the working of the sun compass—first discovered in bees by VON FRISCH (1950) and in birds by KRAMER (1950).

Other experiments (PARDI and PAPI, 1953) were aimed at a more quantitative analysis of this unique compass by testing the accuracy of its mechanism. Two graphs (Fig. 9a, b) show that there were usually systematic deviations from

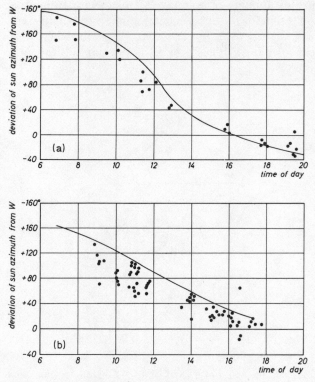

Fig. 9a and b. Sun azimuth curve (solid lines) of June 8 (a) and of unspecified date during the fall experiments (b), 1952 and a number of mean escape directions recorded during May/June and September–November 1952. The scores are plotted so that scores of ideally correct directions would fall on the azimuth lines. From PARDI and PAPI (1953)

the exact direction. Positive phototaxis, i.e. the tendency of the animals to deviate towards the source of light, is probably always more or less superimposed on the compass resulting in some measurable deviation towards the sun. Two major questions were subsequently considered:

(1) How does the sun compass function at greatly differing geographical latitudes?

(2) How do these animals orient at night?

The first question arises because the apparent path of the sun varies greatly with latitude (and also to a lesser extent with season). As questions (1) and (2) are interrelated, they will be discussed together.

1. Sun-Compass Orientation at Different Latitudes. For these experiments, the lycosid spiders (mostly *Arctosa cinera* and *Lycosa fluviatilis*) have been chosen by PAPI and SYRJÄMÄKI (1963), because, apart from exhibiting good sun-compass orientation (PAPI, 1955b; 1959), they are distributed in Europe over a wide latitudinal range from northern Finland to the Mediterranean.

The experimental methods applied to spiders were different from those used with sand hoppers. The spiders were placed, singly, in a dish placed on water (Fig. 10). Riding on the surface the fast-moving spider frequently bounces against

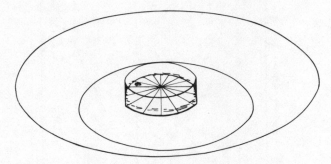

Fig. 10. Experimental arrangement for experiments with *Arctosa cinera*. The spider was placed on water into the dish. View of landmarks was excluded by a circular screen. During an experiment the spider repeatedly bounced against the wall of the dish attempting to escape; the direction of these attempts was recorded by direct observation. After PAPI (1955b)

the periphery with some intermittent pauses, the escape direction coinciding with the direction leading from the water to the shore in the individual's native habitat. The direction of each new successive escape attempt is recorded as one score, a testing time of 2 min usually yielding a few dozen scores from each animal. It should be remarked here, that these scores are even less independent and less suitable for subsequent statistical treatment than those obtained in experiments with *Talitrus* (p. 4).

Italian specimens of *Arctosa cinerea*, while correctly allowing for that portion of movement of the sun that they experienced in their normal habitats (Fig. 11 black scores), were incapable of a definite orientation when translocated to Arctic Scandinavia at 69°01′ N and 20°51′ E and exposed to the sun at times which they had never experienced (not illustrated). The result was confirmed in an experiment in which time-shift was used rather than latitudinal translocation (Fig. 11, white scores). In this approach the internal clock was shifted by 12 h to a reversed day, their physiological "night" now roughly coinciding with the natural day in Italy, so that the experimental animals could be exposed to and tested under the sun there. If properly oriented, the white scores should follow the solid line through the "night" time as the black scores follow the dotted line during day time. Several *Arctosa* populations of the same species and also of the related *Lycosa fluviatilis*, which naturally experience the midnight sun, however, were properly oriented throughout 24 h (not illustrated).

One major complication of the sun compass occurs in the tropics. There, the sun moves clockwise and culminates in the south for half of the year;

9

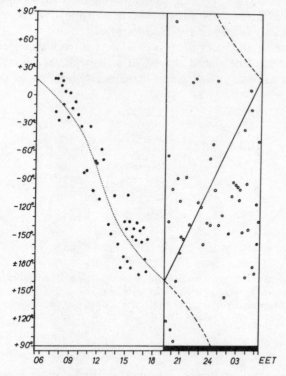

Fig. 11. Sun-compass orientation of an Italian population of *Arctosa cinera* throughout 24 h (abscissa in eastern European time). The escape direction was east. Each score represents the mean direction of the escape attempts performed by one spider within 2 min. Properly oriented during daytime the black scores follow the dotted curve representing the theoretical mean in the left part of Fig. The white scores (right part of Fig.) were collected during night-time as explained in the text. From PAPI and SYRJÄMÄKI (1963)

during the other half it moves counter-clockwise and culminates in the north. PARDI and ERCOLINI (1966) have studied tropical populations of *Talorchestia martensii*, a relative of *Talitrus saltator* which lives on East African beaches. These animals were able to orient correctly during both periods of solar culmination and direction of movement (clockwise and counter-clockwise). However, animals translocated suddenly from their native habitat (where the sun was culminating north at the time of the experiment) to a region where the sun was culminating south, were unable to switch as they do under natural conditions in the course of the year. These experiments demonstrate ecological adaptations to local solar conditions.

2. Orientation at Night. The previous Section has discussed sun-compass orientation in spiders extending, partly for methodological reasons, into the animals' night, in which at least those from southern latitudes are not particularly active. In the sand hopper *Talitrus* the situation is different since these animals are

10

particularly active at night and need some means of guidance, certainly for their nocturnal inland excursions (p. 2, 3).

For nocturnal experiments the technique had to be altered. First of all, seaward movement could only be occasioned when the experimental jar was heated and the air inside was dehydrated; and secondly, recording was made by vacuum-flash photography (PAPI and PARDI, 1953; 1959; PAPI, 1960). It has already been mentioned that on clear, but moonless nights, no directed movements could be observed. When exposed to the moon, however, two different populations showed two different directions of movement, according to their differently oriented home beaches (Fig. 12). The experiments to segregate the action of the moon as a guiding factor were usually performed as follows:

(1) The mirror experiment yielded results comparable to the mirror experiment with the sun (p. 5), however, only with full or nearly full moon were the animals to any extent well-oriented.

(2) Orientation with an "artificial moon" (flashlight) was observed only during the normal phase of full moon.

(3) If clouds obscured the (natural) moon, disorientation resulted.

These results indicated the operation of a moon compass, analogous to the sun compass. However, explaining nocturnal orientation by a moon compass is much more difficult and complex than is diurnal orientation by a sun compass. (1) The moon's position at given hours of day or night changes much more drastically from night to night i.e. in the course of the year. (2) For extended periods during the year, the moon is visible only or partly during the daytime or (3) is not visible at all during the phase of new moon. Thus, the special conditions of near full moon visible at night, conditions under which oriented

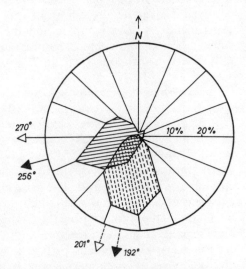

Fig. 12. Percentage distribution in experiments on moon-compass orientation of *Talitrus* from two populations. The expected direction of one population is 270°, the resultant of the experimental distribution (solid line shaded polygon; N = 2143) is 256°; the expected direction of the other population is 201°, the resultant of the experimental distribution (broken line shaded polygon, N = 689) is 192°. After PAPI and PARDI (1959)

11

behavior has been confirmed, are frequently not available and the ecological significance of such a moon compass would be rather limited.

Another complication arises from the fact that *Talitrus* has also been shown to use the sun for orientation, in a special way, under artifically created circumstances, i. e., when exposed to it at night in phase shifts. These experiments demonstrated that *Talitrus* also has an idea at night of a movement of the sun, however unrealistic, in addition to that of the moon. But the movements and positions, respectively, of moon and sun hardly ever agree. In addition to two different clocks, a moon clock and a sun clock, one would, thus, have to presume that the animals distinguished sun and moon perhaps by their qualitative or quantitative difference in light emission (or reflection). Experiments to test this have, so far, not been carried out.

Finally, PAPI and PARDI's findings of moon-compass orientation have been partly confirmed and partly challenged by ENRIGHT (1961) using *Orchestoidea corniculata*, a species related to *Talitrus* and living on Californian beaches. ENRIGHT repeated the mirror experiment and confirmed the basic notion of the moon being used as a compass under certain conditions. Findings in which ENRIGHT and PAPI and PARDI did not agree concern methodological details such as the effects of vacuum-flash photography, recording of time of catch, how to keep the animals before the experiment and, finally, the question of whether a continuously operating lunar clock or a single-cycle night clock, to be initiated possibly by sunset or moonrise, would enable the animals to compensate for the movement of the moon. These questions have not been settled. The orientation of *Talitrus* was again investigated by ERCOLINI and SCAPINI (1974), who demonstrated that the slope of the beach interacts to some extent with solar or lunar orientation seaward or landward.

Even if many details still await explanation by new and more sophisticated experiments, the sun compass and the moon compass may be considered as sufficient explanation of most natural movements and the behavior of experimentally tested arthropods such as *Talitrus saltator*, *Arctosa cinerea* and related species.

In experiments methodologically rather similar to those of PAPI and coworkers with *Arctosa*—and therefore mentioned here—an apparently spontaneous southerly flight reaction of the water skater *Velia caprai* (= *V. currens*) continuing in a dish in captivity was used as a behavior basis (BIRUKOW, 1957; 1960; BIRUKOW and BUSCH, 1975; EMEIS, 1959). Orientation in the day and at night under natural and artificial light regimes etc. has been tested and beautiful curves have been published. The experiments and their results have been widely cited in the field of animal chronometry and sun-compass orientation. But attempts by HERAN (1962) to confirm the findings of BIRUKOW and coworkers failed. *Velia*'s spontaneous flight reaction to the south which had served as experimental criterium in all experiments by BIRUKOW and coworkers could not be repeated and has never been claimed again. It would, indeed, be hard to visualize the ecological significance of such a stereotyped course absolutely in contrast to the ecologically useful flight directions demonstrated in *Arctosa*, *Talitrus* etc. *Velia*'s southerly flight reaction was probably nothing but a supposition introduced and kept alive by improper methodology.

12

2. Locusts

A. Field Performance in Orientation

Migratory locusts never return to locations where they have been before; they do not "home". Their movements are aperiodic and irregular but, nevertheless, they represent some of the most spectacular animal migrations on record.

Considerable discussion is going on as to which movements deserve the term migration and which do not (URQUHART, 1958; JOHNSON, 1969). Rather than tangling with terminology (see Preface) I prefer to select a few outstanding phenomena of long-distance movement.

The desert locust *(Schistocerca gregaria)* is commonly known to perform mass migrations covering up to several thousand kilometers. Individual swarms have been tracked, as for instance by WALOFF (1959) in 1950, when a swarm moved from the Arabian peninsula over more than 5,000 km to the West to Mauretania, in less than two months. In 1951–1952, another swarm took less than 4 weeks to migrate over more than 3,000 km from the Somali peninsula North-north-west to Northwestern Jordan and Western Iraq, two impressive examples which suffice to illustrate this animal's normal migratory capacity. Examples of unusual migration are discussed at the end of the following chapter on experimental and theoretical analysis (p. 14–16). Due to the economic importance of this animal, intensive research has provided valuable insight into its migration habits. A fairly simple explanation has been found for the questions why and how these animals migrate.

B. Experimental and Theoretical Analysis

After passing through the nymphal stages and becoming able to fly, the hoppers take off in large numbers and move or, rather, are moved downwind. SAYER (1956; 1965) and RAINEY (1963) have analyzed directions of individual specimens within swarms and found that they are not uniform, members of groups within a swarm seeming to have the same direction but groups having different directions (Fig. 13). At the edges of a swarm, the animals seem to turn back into it, thus preventing dispersion of the swarm. The general directional movement of a swarm is largely determined by the direction of the wind (RAINEY, 1951; 1963). Moving downwind seems to have considerable adaptive value for the hoppers, bringing them in the regions where they occur—the tropics and adjacent

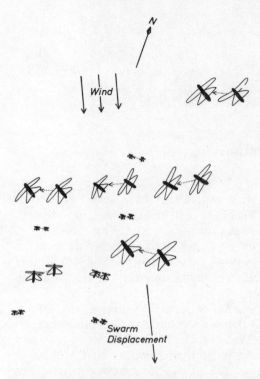

Fig. 13. Movement of two groups of locusts (at different altitudes and therefore of apparently different size) of one large swarm recorded by double exposure photography. The individuals of the two groups had different headings (NW and N), slightly different tracks (dotted arrows) and the swarm had an overall displacement downwind to the SSE. From JOHNSON (1969)

areas—sooner or later into the so-called Inter-Tropical Convergence Zone (ITCZ, Fig. 14a, b) in which, by and large, opposing warm and humidity-laden trade winds meet, air ascends and, in cooling off, precipitates its humidity in the form of tropical rain. Rain is exactly what locusts need for reproduction, their eggs having a high demand for water for development. The ITCZ moves seasonally and its movement may at least partly explain the—however aperiodic—movements of locusts. Thus, the direction of over-all movements is hardly controlled by the animals themselves. Nevertheless, extrapolating from a large number of experiments with other insects and as suggested by KENNEDY (1951) it may be safely assumed that locusts also possess the sun compass and that they probably use it to maintain headings. There is, however, evidence that locusts also migrate in moonlight (HASKELL, 1966) indicating that more than one orientation system is involved. There is no evidence that locusts are able to navigate. With wind being such an important factor in locust migration it is not surprising that, however exceptionally, locusts are drifted by winds to locations where they do not normally occur. WALOFF (1946) reports on such flights beyond the normal locust range including some extending from Africa up to 2,000 km

14

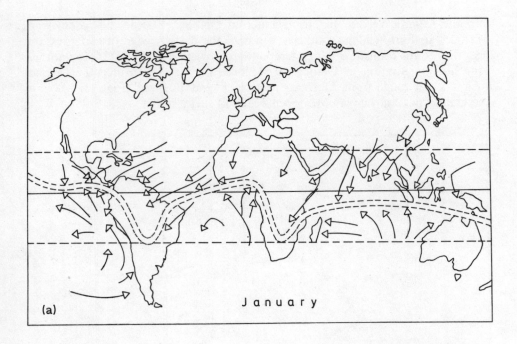

(a)

January

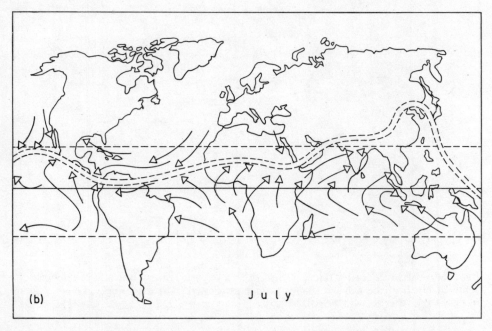

(b)

July

Fig. 14a and b. The Inter Tropical Convergence Zone (broken lines) and the prevailing winds (arrows) at different seasons (a) in January and (b) in July. After JOHNSON (1969)

into the Atlantic Ocean. In one instance in October 1945 locusts rode from Marocco to Portugal, the southerly winds of the eastern edge of a depression near the Azores coinciding with a temporary destruction of the ITCZ. In another very similar situation in October 1954, locusts were even drifted far beyond their normal range from Marocco to Ireland and England (Fig. 15) also for the greater part of the way over open ocean RAINEY, 1963).

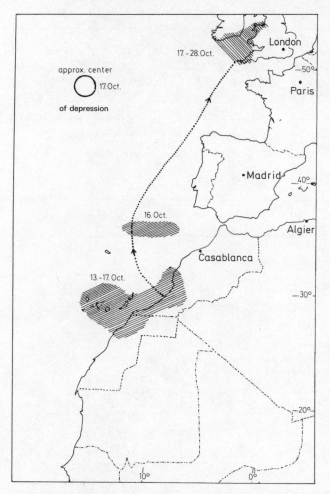

Fig. 15. Probable track of desert locust *(Schistocerca gregaria)* riding the winds of a depression centered north of the Canary Islands from northwestern Africa to England and Ireland in Oct. 1954. After RAINEY (1963)

16

3. Bees

Among arthropods, research on homing in bees is most advanced of all, deserving, therefore, a separate Chapter. First-hand and first-rate reports and detailed reviews on this fascinating topic are commonly available (V. FRISCH, 1965; 1968). However, because of their paramount importance in this field and their stimulation of navigation research in general, the most important aspects are discussed here.

A. Field Performance in Orientation

A naive observer at a bee hive may see bees flying out and eventually coming back loaded with nectar or pollen. Marked bees may be observed in the immediate vicinity of the hive as well as several kilometers away, over low elevations and large obstacles such as mountain ridges. How does a bee find its way home to the hive after an excursion of possibly several kilometers, spent busily zigzaging in search for food?

An observer may also notice that at first in the morning only a few bees arrive at a rich source of food such as a flowering bush or, in an experiment, at a dish with highly concentrated sugar water. Very soon afterwards scores of new collector bees crowd the bush or the dish. The first bees appear somehow to have alerted their hive mates to come out and participate in harvesting. How can the scout bees possibly have communicated the location of the yield?

B. Experimental and Theoretical Analysis

Let us assume that we have set up a dish filled with sugar water and scented with mint oil about 5m from a bee hive, as has been done many times by V. FRISCH and his students. We mark the first scout bee to arrive with a small patch of paint and watch its behavior upon arrival inside the hive through a glass window, and see the following: the scout bee climbs up on the surface of a comb and upon distributing its harvest to other bees, begins to run in a specific, recurring pattern on the vertical comb. The bee "dances" in circles, every once in a while changing its direction of movement (Fig. 16a). Several other bees follow the dancing scout bee; if we also mark them, we soon find them sucking sugar water at our dish outside the hive. This "round dance" seems to communicate to newly recruited bees where to find food near the hive. The scent also has its function; if we set up a few more dishes either without any scent or with some other scent than mint oil they are largely ignored by the new foraging bees. If the experimental dish is moved somewhat away from the hive, the observer at the hive may see a somewhat different dancing pattern, the so-called "sickle dance" (Fig. 16b), indicating to the other bees, who follow in the dance, a somewhat greater distance than the round dance. Upon moving the experimental dish even farther, e.g. 250 m, away, we may record a dancing pattern shaped approximately as a figure eight (Fig. 16c), both halves alternatively run by the bee and its followers. The most obvious

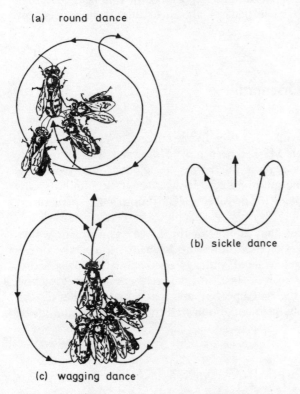

(a) round dance

(b) sickle dance

(c) wagging dance

Fig. 16a–c. Three distinct patterns of bee dance implying specific information, (a) round dance indicating short distances and no direction, (b) sickle dance indicating direction (straight arrow) and unspecified medium distances, (c) wagging dance indicating direction (straight arrow) and specified longer distances. After V. FRISCH (1965)

difference to the other two patterns is the wavy center line of the figure eight. When dancing the center line the bee characteristically waggles its abdomen at the same time producing a jarring vibration with its flight muscles. If the experimental dish is moved even farther away from the hive, no further change of dancing pattern may be observed, however, dancing becomes slower with

18

increasing distance. Thus, the experimenter may estimate the distance by measuring with a stop watch the time of one lap or, more exactly, of the time for one wagging run along the center line of the figure eight (Fig. 17). The newly recruited bees presumably also measure and average the time of several laps or wagging portions when following the dancing scout bee.

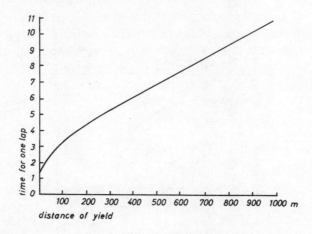

Fig. 17. The distance information in the wagging dance of bees. Relation between time for one lap (ordinate in *s*) and distance between hive and yield (abscissa in *m*). After v. FRISCH (1965)

Not only distance but also direction of a yield may be communicated by the dance. The round dance does not give information on the direction. Yields in the immediate vicinity of the hive may be found quite easily by their scent. If for example four experimental dishes were posted 5 m away around the hive, they are all visited with about the same frequency by experimental bees. Both patterns used to communicate a yield at greater distances, the sickle dance and the wagging dance, also communicate direction. If bees have the chance to dance in the open on a horizontal surface such as the landing board of the hive, the opening of the sickle (cf. Fig. 16b) or the center line of the wagging dance coincide with the direction to the yield. The sun is used as reference, i.e., the angle subtended between the opening of the sickle or the wagging lap of the figure eight coincides with the angle to be taken by the bee on its way to the yield. This means that the bees use a sun compass.

Dancing on the landing board is, however, the exception. Usually, returning scout bees dance on the vertical comb inside the hive in the dark, out of sight of the sun. The bees master this situation in a most astounding fashion: they translate the angle to the sun into an angle to the vertical or to gravity, respectively. The insert Fig. of Fig. 18 demonstrates how a sun angle of 45° left of the sun is "danced" as an angle of 45° left of the vertical line. Newly recruited bees follow the dancing scout bee on the vertical comb and translate the angle to the vertical back into a sun angle.

Thus, direction of the wagging portion communicates direction, speed of wagging supported by vibration signals communicates distance. These are the basic features. Many additional details have been elucidated, giving deeper insight into the entire process of orientation and communication.

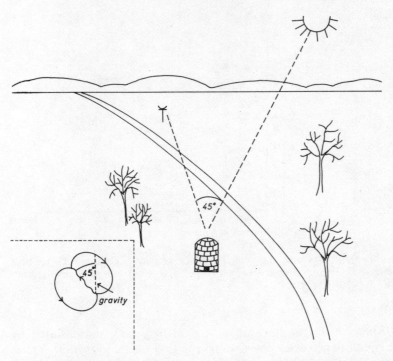

Fig. 18. The directional information in the wagging dance of bees. The angle between the center line (wagging lap) of the "eight" and gravity (inserted Fig.) corresponds to the angle between dish and the sun e. g. 45° in this Figure

Usually, there is at least some wind. Bees cope with cross winds which they sense with sensory hairs on their compound eyes, the antennae, and optically, by advancing their flight-angle to offset the drift. This means they refer to the sun at an angle different from the angle obtaining for the same flight in still air (Fig. 19). Nevertheless, upon return to the hive, they communicate to their mates the angle for still air. This indicates that they notice and allow for wind drift and it means, furthermore, that they calculate and communicate angles with the sun which they have not actually experienced.

Similar abilities are demonstrated by the bees when they fly detours around obstacles rather than the direct route. Their communication system permits only one item of information on the distance and also only one on the direction. A bee scouting around a ridge (Fig. 20) cannot communicate two angles and two distances. However, it is able to extrapolate the direct direction from the two (or more) angles actually flown. It is not able also to calculate the correct

20

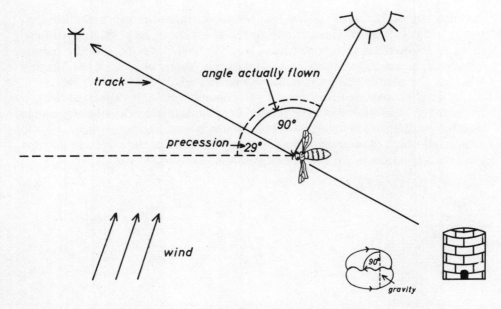

Fig. 19. How bees cope with cross winds. A precession of 29° is necessary to compensate for the wind and to arrive at a dish actually 90° left of the sun. These 90° and not the 119° actually experienced by the bee are communicated with reference to gravity (inserted Figure). After V. FRISCH (1965)

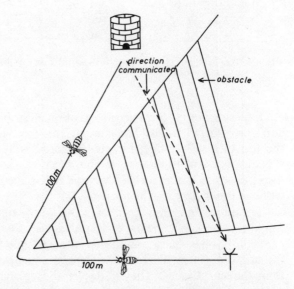

Fig. 20. How bees cope with detours. Bees led around a ridge extrapolate and communicate not the direct distance (broken line) but the actual distance flown (solid line). After V. FRISCH (1965)

distance but simply adds up the two (or more) distances which it measured by the energy expended. Thus, no difficulties arise if the direct route as communicated to followers is in fact nearly as long (up and down hill) as the detour around. If, however, the direct route is markedly longer or shorter the foragers will miss the goal by flying too far or not far enough.

In addition to this restriction of the communication system, bees lack a term for "up" and "down". Under natural conditions this deficiency never causes any trouble. If a yield is up or down hill—which is very commonly the case—the forager informed by a scout simply follows the slope in the direction and for the distance indicated. An extreme situation is illustrated in Fig. 21.

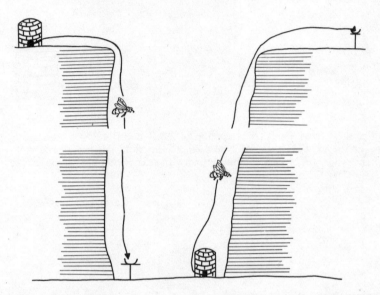

Fig. 21. An extreme situation of down hill and up hill and how the bees cope with it. After V. FRISCH (1965)

Experimentally, however, a situation may be created revealing the lack of "up" or "down" in the bee's language: if a yield is moved step by step to the top of a tall bridge, foragers alerted by a scout bee and informed as to the distance and direction of the yield will follow the slope in the correct direction and, if the scent does not reach them search for it on the ground without flying up into the air (Fig. 22).

The directional mechanism of the bees is the sun compass. The sun is frequently not directly visible, being obscured by a cloud, a mountain or some other obstacle. Under these conditions, a small patch of blue sky suffices for the orientation system. Then the bees use the polarization pattern of the sky rather than the sun directly. They allow for the diurnal variation of the polarization pattern just as they allow for the sun's diurnal movement.

One phenomenon puzzled V. FRISCH and his students for years: the direction indicated by the bee's dance inside the hive almost always included some error.

22

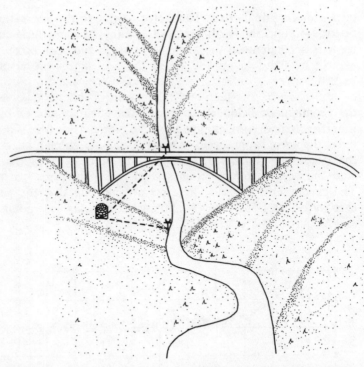

Fig. 22. An experimental situation in which the lack of "up" or "down" in the bees' communication system is made apparent. Foragers informed by scout bees on distance and direction of the dish on top of the bridge will not fly up through open air but arrive at the control dish under the bridge. After v. FRISCH (1965)

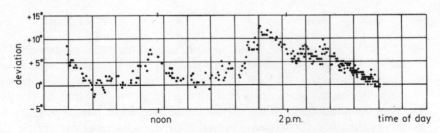

Fig. 23. An example of the error in the bees' dance (deviation from the correct angle to gravity, ordinate) in the course of a day (abscissa). Each dot indicates the average of 10 measurements in one dance. After LINDAUER and MARTIN (1968; 1972)

An example is given in Fig. 23. Deviations from the correct angle range up to more than 30°. Nevertheless, alerted foragers reached the goal without corresponding directional deviations. Thus, the error made during translation of the sun angle to a gravity angle was eliminated during re-translation of the gravity angle to a sun angle. LINDAUER and MARTIN (1968, 1972) finally clarified the mysterious residual error: The "error" is due to the interaction between gravity orientation and the earth's magnetic field. The magnitude and direction of the error is dependent on the angle with which the lines of the geomagnetic field are cut by the dancing bee, i.e. if the bee dances on a vertical comb facing east, the error is different from that of a bee dancing on a comb facing, for example, south. The temporal fluctuation of the error corresponds to that of the geomagnetic field. If the geomagnetic field is altered by Helmholtz coils, the error changes correspondingly; if the geomagnetic field is cancelled experimentally, the error approaches zero (Fig. 24). It should be noted at this point that these experiments demonstrate an interference of the geomagnetic field with the bees' communication system (which involves orientation to gravity). The experiments reported here do not indicate that the geomagnetic field is used for orientation. A review of these findings has been published by LINDAUER and MARTIN (1972).

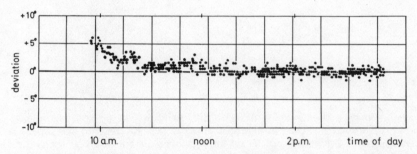

Fig. 24. The error in the bees' dance diminished after the geomagnetic field was cancelled by Helmholtz coils. Symbols as in Fig. 23. After LINDAUER and MARTIN (1968; 1972)

The same means of orientation and communication as described in the preceding Sections for a food-source are used by swarming bees to indicate a suitable new home. At least part of the dancing is then performed on the cluster in the open, components of the dance giving information on the distance, the direction and the quality of the prospective new home. If two or more seemingly suitable places have been found, dancing and inspecting goes on until an agreement has been reached as to which of them is the best. This may take several days, involving changing weather conditions and testing the suitability of the new home in rain or hot sun. Once an agreement has been reached, the swarm moves to the new home. If this should take too long the swarm may perish.

The analysis of the orientation system of bees being thus far advanced, and our knowledge of it being far ahead of that of other animals, a few theoretical considerations may be justified.

The navigation system of bees may be classified as vector navigation, comprising the two parameters, distance and direction. A vector is a quantity defined by direction and length (=distance).

24

By following a dancing scout bee, newly recruited bees obtain two sets of information: one as to the distance and another as to the direction of a source of food. If displaced en route or transported away from home by an experimenter as is usual, for example, with homing pigeons, bees are unable to allow for the displacement, and will not home. This deficiency makes vector navigation a system of limited scope and clearly distinct from a system of true navigation. True navigation would enable an animal to home upon displacement from unfamiliar territory. It is a matter of definition whether or not vector navigation qualifies as navigation. It may be called vector orientation—a more general term.

Scout bees searching for food utilize a system that may be called route reversal, also comprising the parameters distance and direction and integrating vector navigation. On its way out, a scout bee integrates the directions actually flown into the direct direction to home and adds up the distances covered. From the point of return it has only one distance and one direction to fly and, possibly, to communicate to its hive mates.

4. Butterflies

A. Field Performance in Orientation

One of the best known migrants among butterflies is the monarch *(Danaus plexippus)*. This beautiful butterfly inhabits parts of the North American continent. Its existence depends upon the availability of the milkweed *(Asclepias* spec.) the exclusive source of food of the caterpillars. Milkweed thrives from Canada down to the Gulf States and on the West coast into Mexico. The monarch migrates roughly within the range of this plant. Monarch migration may be as spectacular and impressive to a field observer as is bird migration. Day after day in late summer and fall, the air is full of monarchs wherever one

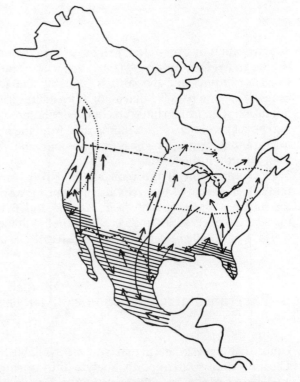

Fig. 25. Summer range (enclosed by dotted lines) and winter range (horizontally shaded) of the monarch butterfly *(Danaus plexippus)* in North America. Assumed migration routes are given as arrows

looks, and it is clear that a continuous stream of very large numbers of butterflies is moving in southerly directions. Migrating monarchs cover the entire eastern part of the United States, the Mississippi valley and move also along the Pacific coast (Fig. 25).

Marked monarchs have travelled record distances of over 3,000 km. Examples taken from labeling records are given in Table 1. Up to 128 km daily performance has been calculated by URQUHARDT (1960).

Table 1. A few examples of distances covered by monarch butterflies *(Danaus plexippus)* marked in Ontario during southward migration. (From JOHNSON, 1969 after URQUHART, 1960)

Marked (Ontario)	Reported	Distance (km)
July 2	Oct. 18, Texas	2,124
Sept. 3	Oct. 18, Texas	2,124
Sept. 12	Oct. 9, Mississippi	1,706
Sept. 13	Oct. 25, Texas	2,164
Sept. 18	Oct. 5, Mississippi	1,706
Sept. 18	Jan. 25, Mexico	3,009
Sept. 19	Dec. 27, Texas	2,254
Sept. 30	Nov. 7, Texas	1,984

These findings very clearly indicate that the entire distance may be covered by the individual. BROWER (1961) reports extensive breeding of monarchs in Florida in winter and the same is probably true for other parts of the wintering range. It is quite possible that all or at least large fractions of the vernal northward migration stem from winter-breeding populations rather than from overwintering individual specimens which immigrated from the north in the fall. The northward migration in spring is much less obvious and much less is known about it than about fall migration.

The migration of the monarch in North America is but one example of butterfly migration. Monarchs also migrate in Australia, and several other species, e.g. members of the genera *Ascia, Vanessa, Aglais,* and *Pieris* have been reported to migrate over at least several hundred kilometers in various parts of the world. For detailed references see e.g. JOHNSON (1969).

B. Experimental and Theoretical Analysis

Unfortunately, very few actual facts are available for a discussion in this chapter. The only major attempt to analyze the monarch's orientational capabilities involved marking and the analysis of marking and recovery records. From these it is evident that individual butterflies may travel long distances one way from the northern breeding range to southern winter quarters. It may be taken for granted that immigrants from the north breed in the south but there is no clear-cut and unequivocal evidence that the northern butterflies return to the northern breeding range next spring. According to URQUHART (1960), the physical condition of at least some portion of spring migrants arriving in northern latitudes proves that they are individual specimens that have overwintered. However, spring recoveries of individual specimens marked in California revealed only relatively short-distance movements (about 400 km to the north). The idea that short-distance movements of successive generations account for the overall migration pattern in spring cannot be dismissed at the present time. Accordingly, ability of only directional orientation—compass orientation—may largely suffice to explain the migration of the monarch and probably of other insects. Even if individual specimens should be credited with completing two or more migration periods, true navigational ability could not be presumed as long as evidence for return to specific geographical goals is lacking. No invertebrate has, so far, been proved capable of true navigation.

A large number of field observations leave no doubt that butterflies migrate in a persistently straight and unidirectional fashion. Lacking appropriate data it is impossible to assess the extent to which the track is controlled by the animal, especially over long distances, in cross winds, in changing wind directions and in other varying conditions. Evidence that butterflies may migrate by day, night and under overcast skies suggests, once more, possibly redundant compass systems even for the rather simple task of directional orientation.

Thus, much remains to be done, preferably in terms of quantitative field observations and, especially, of controlled experiments, in order to explain the bases of butterfly migration.

5. Fishes

Although most reviews or books on "fish" migration refer exclusively to teleosts, outstanding migrants are to be found also among the cyclostomes and elasmobranchs. Their movements are very similar, if in fact they differ at all, as far as motivations and—probably—modes are concerned. The methods employed to investigate these migratory movements are also more or less the same. Despite their phylogenetic differences, for the purpose of this book the members of three different classes of vertebrates will be discussed together, as members of different classes of arthropods have been discussed together in previous chapters for the same reasons.

A. Field Performance in Orientation

Movements of fishes are widespread, complex, and of many different kinds and modes, very similar to, but much less obvious than those of bird migration. Among the most accomplished and most popular fish migrants are teleosts, the eel *(Anguilla anguilla)* and several salmon species (*Salmo* spec. and *Oncorhynchus* spec.), but others also have impressive migration records.

The Eel. According to HARDEN JONES (1968) 16 species of eels are known, 14 of which are found in the Indo-Pacific region. In the Atlantic region, two species have been established (SCHMIDT, 1906, 1913, 1932) on the basis of different vertebral counts, the American eel averaging 107, $s = 1.2$, the European eel averaging 115, $s = 1.3$ vertebra (EGE, 1939), and on the basis of their geographical distribution. According to the work of SCHMIDT (e.g. 1932), eel larvae hatch in the Sargasso Sea between 22°–30° N; 48°–65° W. Those of the American species drift with water currents to, and northward along, the east coast of North America; those of the European species drift east-northeast with the Gulf Stream. Upon arrival at the continental shelf within one (America) to three (Europe, Africa) years the larvae ("leptocephali") metamorphose into glass eels. As elvers they then move upstream in rivers, the American eels in rivers in the West Indies, the Guyanas, Panama, the United States and northward up to Greenland; the European eels in rivers in North Africa, in the countries bordering the Mediterranean, the Black Sea, the entire European Atlantic coast up around the North Cape and including the Baltic Sea and Iceland (Fig. 26). The eels grow in fresh water for 5 to 9 years. On the approach of sexual maturity they migrate downstream, and out to sea. From here on, hypotheses

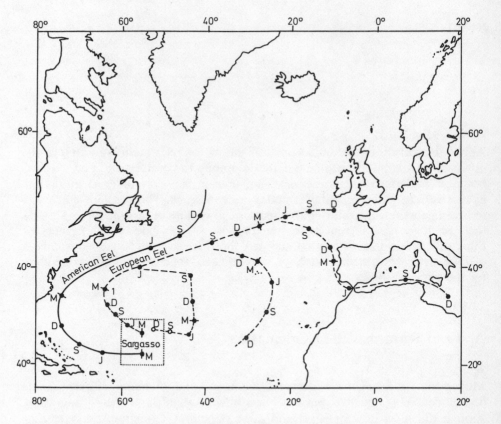

Fig. 26. The journey of eel leptocephali from the Sargasso Sea where they hatched in March (M) plotted in three-month intervals; June (J), September (S), December (D). The end of years is indicated by ♦ and numbered consecutively. After HARDEN-JONES (1968)

replace evidence since no adult eel has ever been caught in the open Atlantic. Nobody seems to question that the American eels do in fact arrive in the Sargasso Sea for spawning. However, for the European eel there are two alternative ideas. SCHMIDT hypothesized that the adult European eels migrate to the Sargasso Sea for spawning, implying that European and American eels are, indeed, two different genotypes. TUCKER (1959) hypothesizes that European eels never succeed in returning to the spawning ground but perish somewhere between Europe and the Sargasso Sea after leaving the coast. Only the American eels, having a much shorter distance to travel, succeed in reaching the Sargasso Sea and in reproducing there. Eggs of the American eel (mean vertebral count 107) will produce American eels only when the eggs experience a sudden temperature change during some sensitive period. If this shock effect does not occur, European eels (mean vertebral count 115) are hatched. For this temperature change to occur or not to occur a close relation to currents to which the eggs are initially exposed would have to exist. Future evidence will sustain one of the two hypotheses, either by catches of European eels en route to the Sargasso Sea or by

32

recovery of eel otoliths from the deposits of the Sargasso Sea, provided there is a difference in otoliths in the two species as suggested by HARDEN JONES (1968). In any case, the portion of an eel's life of main interest here, namely that involving homing, in this case to the Sargasso spawning grounds from the coast of America or even from Europe, is completely obscure as is also the mode of orientation on which this fish may rely. The long-distance trans-Atlantic movement of larvae is passive and does not require means of navigation. CREUTZBERG (1961) has suggested that the eel larvae detect a chemical substance in the ebb water which would have to be common to all European waters attracting the larvae into the estuaries. Orientation near and in the estuaries and up the rivers is on the basis of orientation with respect to currents (CREUTZBERG, 1961) i. e., also of non-navigational character.

Of course, no evidence can exist to show whether young eels return to their ancestral river or some specific area which their parents left years ago.

The Salmon. In contrast to the eel, salmon breed in fresh water but most of them spend the greater part of their lives in the ocean. Only a few exceptional stocks of land-locked salmon, such as in the Canadian lakes, are known (RICKER, 1938). Due to the economic significance of the Pacific salmon (*Oncorhynchus* spec.) its life history has been studied more intensively and is much better understood than that of its European relative (*Salmo* spec.) and will be primarily dealt with here. Salmon spawn in fall and winter in oxygen-rich lakes or streams. The hatchlings grow through various stages, named as "fry", "parr" and "smolts", and migrate downstream in spring and summer. Young Pacific salmon emerging from streams and rivers of the American Pacific coast seem to head northwest or west (those from the west coast of Alaska going south), to lead a pelagic life feeding heavily and growing as may be seen from Japanese fishery records (HARDEN JONES, 1968, reviewing papers by ANON and by BROADBENT).

Where the Atlantic salmon spend their time at sea is much less well-known. According to ALM (1958), some salmon from tributaries of the Baltic have been found in southern parts of the Baltic Sea. Salmon from other areas of Europa have been found in the northern North Sea (BALMAIN and SHEARER, 1956) except, perhaps, those from Scottish spawning grounds which have been found between Greenland and Spitzbergen.

After two or more years, occasionally after one year, mature salmon return to fresh water to breed and die (Pacific salmon), or to survive for two or more spawning seasons in subsequent years (some Atlantic salmon). The pelagic portion of the Pacific salmon's migration cycle may cover distances of a few thousand km; the fresh-water portion may, again, cover more than 2,000 km (up the Yukon River, for example) (HARTT, 1972; CLEAVER, 1964; NEAVE, 1964).

There is good evidence that a high proportion of returning salmon finds and ascends its native river in which it hatched years before and there is only slight evidence of straying even to immediately neighboring rivers. This is the reason why these migrations from the feeding to the spawning grounds are those most exciting to the student of fish navigation. Evidence for the return to the native river stems mostly from marking of young salmon migrating downstream and their recovery in the same river or its estuary in later years.

Many such experiments have been carried out on Pacific salmon from California to Alaska. Furthermore, eggs have been released into streams in attempts to stock regions which were, although ecologically suitable, previously not inhabited by salmon, some even as far away as New Zealand. At least some of these transplantation experiments have been successful in establishing new populations of salmon and, presumably, establishing new patterns of migration.

Less than 10 % of all emigrating salmon are estimated as returning at all, the rest being lost at sea. This finding certainly raises the question how many of these other 90 % might have fallen victim to navigational errors. While this question cannot be answered, some more figures are available to illuminate the problem. Tagging of mature salmon in the Pacific (e.g. in the Gulf of Alaska) revealed an estimated overall 10 % recovery rate of fish in estuaries, rivers and streams (HARDEN JONES, 1968, reviewing unpublished results from Biological Station, Nanaimo, Canada). Recovery rates were the higher, the closer to the coast mature salmon were tagged. Survival, thus, is certainly related to the distance to be travelled. Furthermore, chartings by NEAVE (1964) showed that in the Gulf of Alaska, maturing fish separate from the immature fish; shift of the mature fish towards the Canadian Pacific coast has also been recorded. Although these few findings are still far from excluding other possibilities, they at least suggest that directed migration rather than random wandering or drifts lead the mature salmon from the high seas to the area where its native river empties into the ocean, providing olfactory cues which the animals may pick up and follow upstream to the spawning grounds.

Data for European salmon are even scarcer. While the means of orientation during the many thousands of kilometers of oceanic travel remain unknown, there is good evidence that olfactory cues, specific for the parent stream and imprinted on the young salmon soon after hatching, guide the returning mature fish to the ancestral spawning grounds, once recognized probably somewhere more or less near the estuary of that river or stream.

Other Fishes. Apart from these "classical" examples of fish migration and homing, there are many more similar phenomena now becoming known, some—including an example of elasmobranch migration—to be briefly reviewed here. (For more detailed information see, for example, HASLER, 1966; HARDEN JONES, 1968).

Regular long-distance migrations have been described for example for the dogfish *Squalus acanthias*, a small shark. In contrast to eel and Pacific salmon, journeys are repeatedly performed by the individual specimen. The summer feeding grounds of this shark have been found to be in the zone where Gulf Stream and Labrador Stream mix. In fall, supposedly triggered by decreasing water temperature caused by the first fall storms and invasion of water of polar origin, groups of this shark migrate about 2,500 km to the wintering (and mating) grounds off Virginia and the Carolinas. From February on, ♂♂ and ♀♀ return to the feeding range. Pregnant ♀♀ move at about the same time into warmer and more shallow water, give birth and follow the other adults up north, the young sharks following later. The migratory movements of this and possibly of other sharks or of predatory teleosts such as the tuna (*Thunna* spec.) may have their cause primarily in corresponding movements of their

34

preferred prey fish such as those described in the following Sections (e. g. shad and herring). Predators follow their prey, the migration of which is, in turn, controlled by the seasonal availability of food or of suitable environmental conditions for reproduction, which in turn is correlated with water temperature.

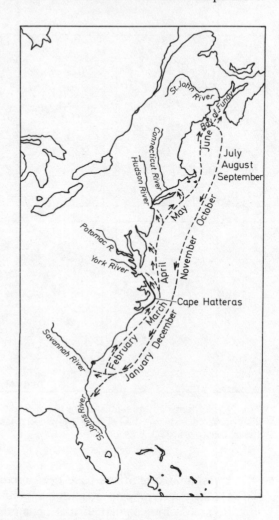

Fig. 27. The migration route of the American shad *(Alosa sapidissima)* in the course of the year and some of its major spawning rivers. From Leg-Get (1973)

Similar to the movements of the dogfish is the migration of the American shad *(Alosa sapidissima)*, as outlined in Fig. 27. The main difference is, however, that the shad enters coastal rivers to spawn. The shad starts out north in January off northern Florida and Georgia reaching the Bay of Fundy in midsummer, some proportion entering rivers on the way north to reproduce. The seasonal movements have been found to be correlated with water temperature (Fig. 28), the shad apparently preferring temperatures between 13° and 18 °C.

Less extensive travels are even more common. Some herring of the North Sea, spawning off the east coast of Scotland, seem to follow a counter-clockwise migration circle (Fig. 29), however, there are many other spawning grounds such as the Dogger Bank, the English Channel and parts of the coast of Norway. As with birds, each population apparently has its wintering ground and migration routes, all of them overlapping each other and making it very difficult or impossible to separate events without reliable tagging.

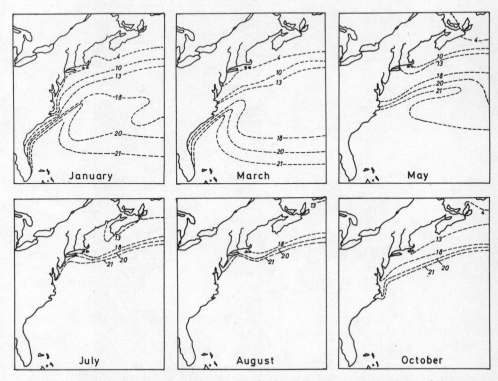

Fig. 28. Isotherms of the water along the U.S. Atlantic coast at certain times of the year. The occurrence of the shad (Fig. 27) is closely associated with the range of 13° to 18°C. From LEGGET (1973)

Very similar migrations have been reported of cod *(Gadus morrhua)* and plaice *(Pleuronectes platessa)*. Some cyclostomes (e.g. *Petromyzon marinus*) migrate in a fashion similar to that of salmon. The young hatch in fresh water and live there as larvae for up to several years buried in the sand and filtering water for food. Eventually, they metamorphose and migrate to the ocean to spend the better part of their life in salt or brackish water, now preying on other fish. They return to rivers and streams, sometimes in impressive numbers, to spawn. There are, however, some cyclostomes (e.g. *Lampetra fluviatilis*) that remain all their life in fresh water—as do some salmonids.

These were but a few examples of fish migration. Some can truly be called homing, even those migrations to spawning grounds which are performed by each individual specimen only once (eel, salmon). Other long-distance travels are merely accomplished by passive drifts (early stages of eel, cod larvae, herring)

36

and do not require or qualify as actively directed movements. Some movements of adult herring after spawning are also considered to be accomplished largely by drift (HODGSON, 1934). HARDEN-JONES (1965) suggests that European eels might get back to the Sargasso Sea by way of a deep counter-current to the Gulf stream, which might at least account for the distance accomplished if

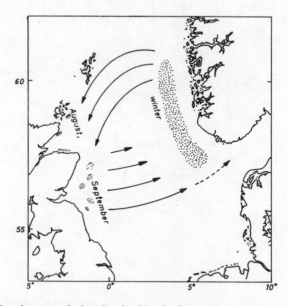

Fig. 29. Seasonal movements of a herring population in the North Sea with spawning grounds (shaded) winter range (dotted) and supposed direction of migration. After HARDEN-JONES (1968)

not for the orientation of movement. No data exist to support this view. Some forms of homing described here imply return not to a specific locality but to some area offering suitable conditions for feeding, wintering or spawning, thus requiring guidance systems with perhaps not so high a degree of accuracy as those for pinpoint navigation.

B. Experimental and Theoretical Analysis

The Eel. Beginning again with the eel, experimental displacements of tagged fish at the North Sea coast of Germany and Holland by DEELDER and TESCH (1970) revealed that some eels, indeed, return to the place of catch from somewhat over 200 km primarily against the general SW–NE current (Fig. 30a). More returned over shorter distances (Fig. 30b). If olfaction was impaired by various (though not always reliable) modes of blocking the olfactory system, homing

37

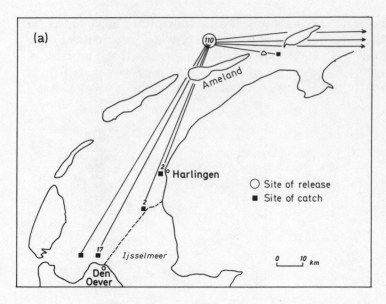

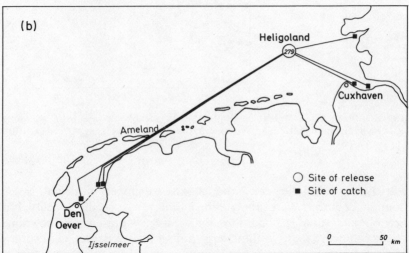

Fig. 30a and b. Results of displacement experiments with tagged eel (a) from Den Oever (Holland) to Ameland (87 km) and (b) from Den Oever to Heligoland (237 km). The numbers in the various symbols explained in inserted Figs. indicate the number of animals involved if more than one. After DEELDER and TESCH (1970)

rates from 8 km to 70 km of distance of experimental animals were no different from those of control animals, a total of 1,057 being involved (TESCH, 1970). Modern tracking methods are now being employed for a more detailed analysis of eel movements (TESCH, 1974). These results and other evidence do not support olfaction as a possible explanation for the established orientational performance of eels.

The Salmon. The notion that salmon recognize their parent stream olfactorily is rather old (BUCKLAND, 1880) and the evidence to support this idea, both regarding the acuity of smell in fishes and from field observations is very good. Salmon home almost exclusively to their parent stream. In a series of laboratory experiments HASLER and WISBY (1951) showed that bluntnose minnows *(Hyborhynchus notatus)* could be conditioned to discriminate waters from different streams previously sampled and brought into the laboratory. WISBY (cited in HASLER, 1966) was also able to condition young salmon to synthetic decoy odors such as morpholine at extremely low concentrations (one part per billion). The experimental apparatus used in this type of experimentation is shown in Fig. 31. Its design clearly illustrates the way it operates. IDLER et al. (1961)

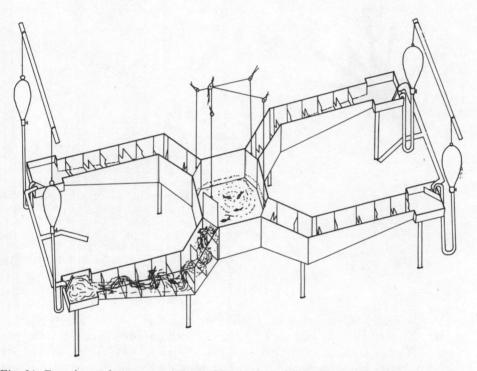

Fig. 31. Experimental apparatus to test the reaction of fishes (primarily salmon) to four specific odors released into each of one of four runways. From the central compartment e.g. salmon fingerlings proceeded upstream against the odor they selected. From HASLER (1966)

have extended similar experiments to sockeye salmon to support the hypothesis that they probably reacted to some chemical substance(s) of the water of what was probably their parent stream. Water from the supposed home stream was added to the experimental laboratory tank. The shoals dispersed and the fish swam faster. No such reaction was observed if equal amounts of water from other streams were added.

39

Both HASLER and WISBY (1951) and IDLER et al. (1961) found that the chemical substance characterizing the parent stream was volatile and heat-labile. Field evidence suggests that young salmon upon hatching in their parent stream are imprinted with its specific odor and specifically react to it years later as mature fish when homing to spawn.

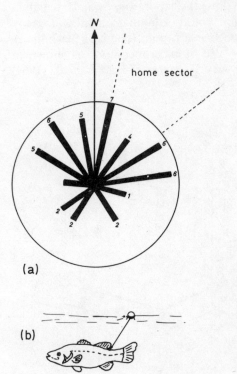

(a)

(b)

Fig. 32a and b. (a) Bearings of white bass *(Roccus chrysops)* corrected for drift one hour after release. The figures indicate the number of fishes in each 11.5° sector. The animals had been taken from their spawning grounds and released in the center of Lake Mendota, Wis. 2–3 km to the SW of the spawning grounds. (b) The animals towing small floats were tracked visually from a boat. After HASLER et al. (1958)

More recently OSHIMA et al. (1969a, b) recorded the electrical response in the olfactory bulb of salmon to waters of various origin, infused into the nostrils. The largest electroencephalographic response was always evoked by the home water but there were also weaker responses to waters sampled along the fresh-water migratory routes of the experimental fishes. These findings confirm the olfactory basis of orientation and they suggest, in addition, that salmon not only recognize the water of the spawning ground but also odors encountered en route.

The remaining problems concern orientation during thousands of kilometers of open-sea travel. For an analysis of the orientation mechanisms it would be important to obtain an answer to the very simple question whether the paths of fish to some goal such as the coast or towards some estuary are straight, regular in some fashion (meandering, spiraling etc.) or random. Since it is extremely difficult to track fish at all, only short-distance movements have been studied. Two classical attempts may serve as examples.

40

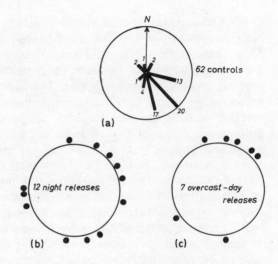

Fig. 33a–c. Mean directions of parrot fish *(Scarus guacamaia* and *Scarus coelestinus)* visually tracked over usually around 120 m upon displacement to presumably unfamiliar locations near their normal range. (a) Headings of 62 untreated control animals under sun, (b) and (c) as indicated. The figures give the number of scores in each direction. Each dot symbolizes the score of one fish. From WINN et al. (1964)

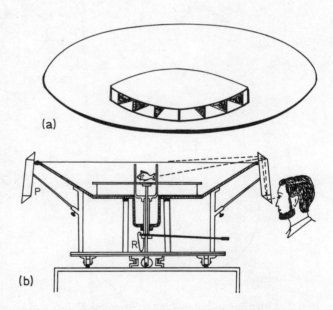

Fig. 34a and b. Apparatus as extensively used by BRAEMER, SCHWASSMANN and others for sun-compass training of fish. Upon release from the center, the fish is trained to seek shelter in one of the 16 boxes. The fish cannot see which box is open until it has passed the edge of the upper platform. After BRAEMER and SCHWASSMANN (1963)

Other Fishes. HASLER et al. (1958) attached small floats to a number of white bass *(Roccus chrysops)* which could be watched from a boat (Fig. 32b), released 55 of them under sunny skies (and 3 more under overcast) in the middle of Lake Mendota, Wis. and tracked them on their supposed way to the spawning grounds 2–3 km to the north east. The distribution of the fish one hour after release is given in Fig. 32a. Although the scatter is rather wide the data are compatible with the view that the fish were on their way to the spawning grounds since the 99 % confidence interval of the sample mean includes the home sector. WINN et al. (1964) caught parrot fish *(Scarus guacamaia* and *Scarus coelestinus)* in their feeding grounds along the south shore of Bermuda and released them at various locations presumably unknown to the fishes. Tracking floats attached to the fish revealed a clear southeasterly heading (Fig. 33a), conforming to the directions fish must select when moving from their day-time feeding grounds to the off-shore caves where they spend the night. When released at night or under overcast skies (Fig. 33b), direction appears to be lost though the number of animals involved is too small for a firm conclusion. Nevertheless, WINN's experiments show more clearly than HASLER's that fishes may move along comparatively straight paths when the sun is visible and that absence of the sun may result in disorientation, suggesting that the sun is used as a compass. Since the sun compass had, in fact, previously been demonstrated

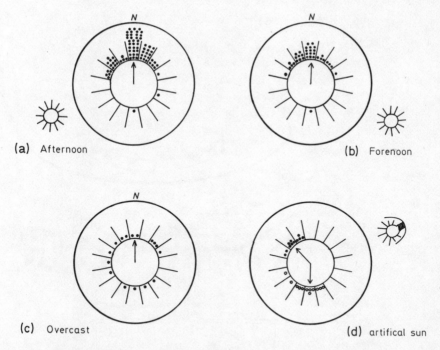

Fig. 35a–d. Classical sun-compass training of a fish. The training direction was north. Each black dot symbolizes one choice i.e. hiding in the box indicated ((a) and (b) under the sun, (c) under overcast skies and (d) with reference to a fixed artificial light). Since this light, in contrast to the natural sun, does not move, the directions of the fish's choice move correspondingly. After BRAEMER (1960)

42

to operate in bees, sandhoppers and birds, laboratory work was aimed at examining the possibility of sun-compass orientation in fish. The case of an aquatic animal is, however, quite different from that of a terrestial animal due to the physical characteristics of water. Under ideal or nearly ideal conditions, i.e. clear water and smooth surface, a fish within a few meters of the surface sees, above him, a manhole with the apparent sun's path in the sky considerably compressed and distorted as compared to the conditions of a terrestrial observer. Nevertheless, the following laboratory experiments show clearly that the sun may be used as a compass by fishes. After some initial attempts employing food-rewarded training the majority of the experiments on sun-compass orientation was carried out with the design pictured in Fig. 34. The fish was trained to seek cover, upon release in the center, in that one of the 16 boxes which pointed in the training direction selected by the experimenter. Fig. 35 provides a convincing account of successful directional training as published by BRAEMER (1960), the sun's role as reference being demonstrated (1) by an apparent breakdown of orientation under overcast skies (Fig. 35c) and (2) by the angles to an artificial light being appropriate for allowance of the sun's apparent movement. The classical mirror experiment with ants by SANTSCHI (1911) has also been performed with fishes (BRAEMER and SCHWASSMANN, 1963; GROOT, 1965). Experiments with young salmon further confirmed the use of the sun compass. But there is no doubt that the availability of the sun may be rather limited in open-sea travel. There is evidence (e.g. GROOT, 1965; FORWARD et al., 1972; WATERMAN, 1972; WATERMAN and FORWARD, 1972; FORWARD and WATERMAN, 1973) that at least some teleost fishes are able to orient to the plane of polarized light. This would allow—as demonstrated in bees (p. 22)—sun-compass orientation under marginal conditions, for example when the sun is not directly visible. But a compass does not suffice for true navigation. Though part of the mystery of salmon homing, for example, is explained by olfactory recognition of home-stream odors, open-sea navigation still remains a mystery.

6. Amphibians

A. Field Performance in Orientation

Many newts, salamanders, frogs and toads have been reported to migrate annually, mostly to and from suitable locations for breeding such as ponds or streams, or between summer and winter hibernation territories. These seasonal movements may be rather inconspicuous, such as those of some Californian newts e. g. of the red-bellied newt *(Taricha rivularis)* as described by TWITTY (e.g. 1959) or by PACKER (1960). The newts live on mountainous and heavily wooded slopes, descending for breeding to selected streams each spring. The migrations may on the other hand be—or become—extremely conspicuous, such as those of the common toad *(Bufo bufo)* as reported by HEUSSER (e.g. 1960; 1964). Due to human interference by road construction, these toads did not only have to cross roads on their way to breeding ponds in spring but also gathered on newly built roads at locations where breeding ponds had been before; they were run over by cars in such numbers that the roads were covered with flattened toads. All authors on amphibian migrations agree that the movements do not exceed a few kilometers, but nevertheless, these comparatively sluggish animals do home to precisely defined goals and the relevant question how this is accomplished needs to be answered.

B. Experimental and Theoretical Analysis

First off all, tagging—mostly by coded toe clipping—of several thousand animals in the course of several years established that individual red-bellied newts, for example, return to precisely the same section of the stream for breeding consistently year after year. If not merely marked and released but also displaced to other sections of the same stream or even across a ridge to some foreign stream up to about 8 km away, a surprising proportion of newts (on the average over 60%) returned to their native stream segments mostly in the next breeding season (TWITTY et al., 1964). In other experiments involving several hundred experimental and control animals blind-folding did not impair homing from up to 3 km (TWITTY, 1966), but upon severance of the olfactory nerves not a single newt with unregenerated nerves returned (GRANT et al., 1968). These results indicate that olfactory cues play at least an important role in homing in newts and probably in other amphibians with similar ecological requirements.

45

In other amphibians with different habits and environmental demands such as, for example, in the upland chorus frog *(Pseudacris triseriata)*, in the southern cricket frog *(Acris gryllus)* or in the Fowler's toad *(Bufo fowleri)* visual orientation may be more important. The example of the common toad gathering on newly built roads exactly at locations where ponds had been before (HEUSSER, e.g. 1960, 1964) also indicates that smell or acoustic signals cannot be the only guiding cues. The use of celestial cues such as sun, stars and moon has been suggested by a series of experiments (e.g. FERGUSON, 1963; FERGUSON et al., 1965; FERGUSON and LANDRETH, 1966; TAYLOR and FERGUSON 1969; 1970). In most experiments, collected animals were released in the center of a circular aquatic test pen of 20 m diameter. The walls excluded the view of landmarks and permitted only a view of the sky and celestial bodies. The animals headed for the periphery and were scored there. The direction chosen was presumably either the direction of their home at the time of the experiment, or was perpendicular to the shore-line on which they lived. Olfactory and celestial orientation was suggested in experiments by TRACY and DOLE (1969a, b) on the Californian toad *(Bufo boreas)*.

Methodology and results were not as clearcut as for example in sun-compass training experiments with fish or birds (p. 42, 56). There is, however, no real reason to doubt the operation of a sun compass in view of the fact that this capacity has been firmly established in so many other animals. It may be more justified to be sceptical on the suggested use of stars or the moon for orientation—if available. Many amphibians move at night as well as under dense undergrowth under overcast skies or even in fog. Other environmental cues may, of course, be involved which have not yet been investigated. Nonetheless, how amphibians home spontaneously or upon experimental displacement is not yet understood.

7. Reptiles

A. Field Performance in Orientation

Looking for a reptile with a record as a long-distance migrant, one inevitably turns to the green turtle, *Chelonia mydas*. It is certainly not the only turtle that covers truly long distances but it is the most outstanding and probably the best-studied migrant. Also known for fairly long-distance movements is the loggerhead turtle *(Caretta caretta)* and related species in tropical and subtropical parts of the Atlantic, the Pacific and the Indian Ocean. Terrestrial turtles cover, if at all, considerably shorter distances.

Like all other sea turtles, adult female green turtles deposit their eggs on the shore in the sand just outside the high-water mark during a short breeding period every third year. Only two breeding grounds of the green turtle remain: a small area on the Caribbean coast of Costa Rica and another on Ascension Island (southern North Atlantic). Upon hatching, young turtles emerge from the sand and go straight across the beach into the sea. Young and adult turtles spend the rest of the time in the ocean. Young turtles disappear for about a year after entering the sea; their whereabouts are a complete mystery. Adult turtles tagged at the Costa Rica breeding site have been reported throughout the western Caribbean, those from Ascension Island have been recovered mostly off the coast of Brazil up to almost 2,000 km away. Individuals return to their specific breeding beach in a three-year cycle.

B. Experimental and Theoretical Analysis

For many years research directed by CARR (1963; 1972) was aimed at obtaining tracks of migrating turtles as a basis for the analysis of the orientation mechanism. In these experiments *Chelonia mydas* and *Caretta caretta* was used. Balloons were attached to animals or to floats towed by animals for visual tracking. More recently, small transmitters were used in the same fashion for radio tracking. Technical shortcomings and the fact that the animals move very slowly—at a rate of only a few kilometers per day—prevented true long-distance tracking. A few sample tracks are given in Fig. 36 and 37 including tracks after experimental displacements over short distances. More complete information may be expected from long-term tracking by satellite which is technically possible today. From these and other tracks it may be concluded that the turtles are capable of

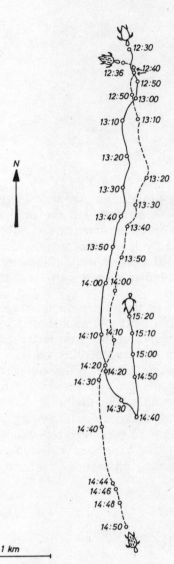

Fig. 36. Tracks of two loggerhead turtles *(Caretta caretta)* with helium-filled balloons attached, on their way to the nesting beach. They were captured when emerging to deposit eggs on the Atlantic coast of Florida and released out of sight of land off the Gulf coast across Florida. Times of day are indicated in intervals. After CARR (1963)

maintaining a fairly straight course. As in numerous other animals this ability may be explained by the use of the sun compass.

From the experimental attempts to demonstrate sun-compass orientation in reptiles those in the terrestrial box turtle *(Terrapene carolina)* by GOULD (1957) had only suggestive results. Those on the most interesting species, *Chelonia mydas* and *Caretta caretta*, by FISCHER (1965) suffered from methodological shortcomings and were not convincing. The same holds true for sun-compass training experiments with lizards (e.g. FISCHER, 1961; BIRUKOW et al., 1963). Although lizards are much handier for experiments than even young marine turtles and are easily trained to search for a warm spot in a circular arena, this neat experimental approach yielded methodologically questionable results.

Despite the lack of actual facts in reptiles, deducing from other animals, especially the closely related birds, reptiles should be credited with the ability of sun-compass orientation. No attempts are known to find an explanation for directed orientation under overcast skies and at night.

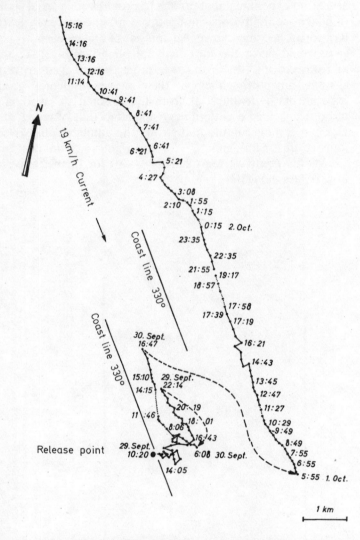

Fig. 37. Telemetered track of a green turtle *(Chelonia mydas)* released at the coast of Costa Rica, caught after laying eggs on its way from the nesting beach. Times of day are indicated in intervals. Dashed lines and arrows indicate where radio contact was lost. After CARR (1972)

Even less is known on how marine turtles navigate. Once again, a compass does not suffice for navigation. KOCH et al. (1969) suggested that some odorous substance emanating from Ascension Island and travelling with the South Equatorial Current westward to the feeding grounds off the coast of Brazil may guide the Ascension Island turtles. To arrive at the breeding grounds, the turtles would have to swim against the current and the odor gradient. This would by no means be simple, and a number of questions arising from such an assumption have not been answered. Apart from attempts to track animals as discussed above, no experimental approach to problems or partial problems such as acuity of olfaction, ability to swim upstream in an apparently featureless environment has, so far, become known. An analysis of the sea currents near the two breeding sites would favor the idea that hatchlings simply have to penetrate the surf and proceed into the open sea in order to be picked up very soon by a surface current transporting them to their known residential grounds. In drift-bottle experiments by CARR et al. from the Costa Rica breeding grounds, recoveries of bottles yielded a pattern very similar to that of recoveries of marked turtles. Upon becoming mature and strong the animals could conceivably reverse the journey upstream "sniffing" for the right smell in the same way as salmon.

At this point we must halt and wait for further experimental evidence on how turtles navigate.

8. Birds

A. Field Performance in Orientation

Most commonly known of all animal migratory movements is bird migration. Banding over more than half a century has revealed summer ranges, migratory routes and winter ranges of many species of birds. Surprisingly many more still remain to be described in comparable detail.

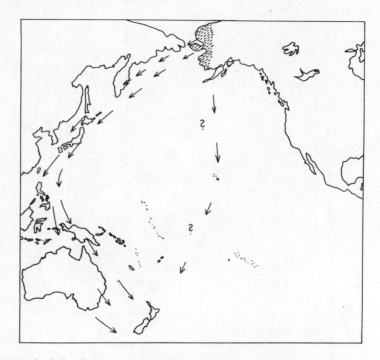

Fig. 38. Breeding range (shaded) of the bar-tailed godwit *(Limosa lapponica)* in western Alaska and its presumed route of migration to Australia and New Zealand

Some birds cover extremely long distances and cross several continents such as, for example, certain populations of the bar-tailed godwit *(Limosa lapponica)*. The populations breeding in western Alaska initially migrate to the southwest along the Pacific coast of Asia, then turn south and southeast across the Malayan Archipelago the Australian continent and winter in New Zealand (Fig. 38). Different geographical populations of other birds have diverging migratory routes;

some converge again to some degree after detouring a migratory barrier; others continue diverging into very different wintering grounds. A well-known example of the former is the white stork *(Ciconia ciconia)* in central Europe. Populations breeding roughly west of the Elbe river head southwest in fall and enter Africa

Fig. 39. Migratory routes of the eastern and western European populations of the white stork *(Ciconia ciconia)* around the Mediterranean. After SCHÜZ (1971)

by way of Gibraltar; populations living east of the Elbe river head southeast and enter Africa *via* the Bosporus and Asia minor, both populations wintering in western, central or southern parts of Africa (Fig. 39). The American golden plover *(Pluvialis dominica)* represents the other pattern. Populations breeding in northern Siberia (Fig. 40)—and these populations extend into western parts of Alaska—head southwest to south, may cross vast bodies of the Pacific and winter anywhere from the Hawaian archipelago to southern Asia or Australia. The eastern population, breeding from Alaska eastward, heads southeast in fall, migrating through the North American continent and wintering in east central South America. The winter ranges of these two populations of the golden plover and possibly those of immediate neighbors in Alaska are extremely far apart.

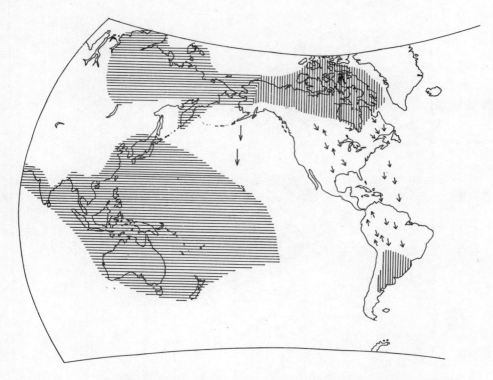

Fig. 40. Breeding ranges and wintering ranges of the golden plover *(Pluvialis dominica)*. Populations from Siberia which extend into western Alaska head south to southwest, the eastern populations head east to southeast in fall. The winter ranges of neighboring populations in Alaska are extremely far apart

Much less spectacular but very considerable migratory performances could be reported from numerous smaller birds such as, for example, the tiny ruby-throated humming bird *(Archilochus colubris)* which migrates across the Gulf of Mexico, a comparatively long distance without a chance to rest or feed (Fig. 41). Many more examples of most remarkable migratory performances could be listed and an even longer tally of short-distance movements could be added such as that of certain populations of the starling *(Sturnus vulgaris)* given in Fig. 42 and discussed on p. 55.

Of the many interesting aspects revealed by banding and by field observations a few should be particularly emphasized here. Contrary to common belief, bird migration only exceptionally proceeds on the north-south axis. Southwesterly or southeasterly directions prevail and westerly and easterly directions for at least parts of the entire journey are known as well. Many migratory routes are not straight but either require changes of headings or are even looped, i. e. the birds use different paths on the way to and from the wintering ranges.

There is overwhelming evidence in favor of pinpoint navigation. Adult birds may return from migration year after year to the same territory or even nest

Fig. 41. Breeding range (dashed lines) and winter range (solid lines) of the ruby-throated humming bird *(Archilochus colubris)*. Many migrants cross the Gulf of Mexico or parts of the Caribbean

site that they occupied the year or the years before. Similarly, the same precise winter territories may be occupied by individuals year after year (e. g. CORTOPASSI and MEWALDT, 1965), while the routes between summer and winter ranges are more flexible. A bird from Scandinavia wintering in Spain may migrate through England in one year and along the coast of northern Germany in another year, possibly depending upon meteorological conditions.

An apparently straight route between summer and winter range by no means implies that the birds use only a compass and do not need more complex navigation systems, as can be learned from displacement experiments with starlings. This leads us to the chapter on experimental analysis.

54

B. Experimental and Theoretical Analysis

If birds are not just banded and released but are banded and released after some substantial displacement, results may become much more informative. From many similar experiments, those by PERDECK (1958) on starlings have yielded the most clearcut results (Fig. 42). Starlings breeding in the Baltic region migrate through Holland and winter in southern England, southern Ireland and northern France. PERDECK caught and banded more than 11,000 starlings on their fall migration in Holland and displaced them to and released them in Switzerland. Adult starlings with previous migration experience headed north-west for their ancestral winter range, while young starlings on their first fall migration headed southwest, parallel to their ancestral route of migration and wintered in southwestern France, Spain and Portugal. This differential behavior of young and adult birds signals the operation of two types of orientation. Adults are able to navigate in the true sense of the word i.e. to find a goal from unfamiliar territory without piloting along familiar landmarks; young starlings appear to have information on the direction to take up from the breeding range, a compass and, possibly, some information on the distance to cover. This is a system with limited scope. If the birds are displaced as in PERDECK's experiment by man or in the field, for example, by strong winds, they are unable to locate the goal which would be the ancestral winter range and a new winter range is established.

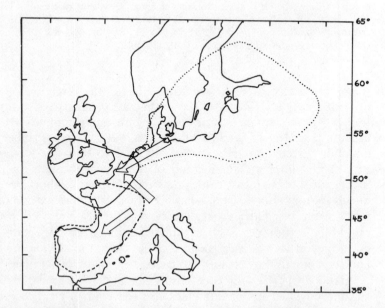

Fig. 42. Breeding range (dotted line) and winter range (solid line) of starlings that migrate through Holland. The other symbols refer to PERDECK's experiments which are discussed in the text. After PERDECK (1958)

Directional orientation as exhibited by young starlings and as demonstrated for example on radar screens by birds flying straight for hours, is easily explained. Birds use either the sun, stars or the earth's magnetic field to keep directions; they have a sun, a star, or a magnetic compass.

The sun compass was first demonstrated by KRAMER (1950) in starlings in experiments based on the spontaneous migratory restlessness of birds and in training experiments (KRAMER and ST. PAUL, 1950). In the experiments with migratorily restless birds a starling was kept in a cylindrical cage. If sunlight was permitted to enter through a window (Fig. 43a), the bird headed WNW.

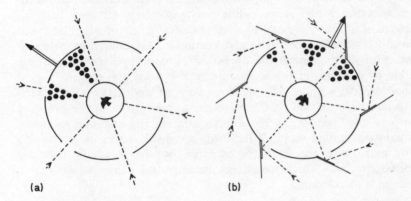

(a) (b)

Fig. 43a and b. KRAMER's mirror experiment to demonstrate the operation of the sun compass. Each black dot symbolizes the mean direction of the bird in a 15-sec interval. The overall mean direction is indicated by an arrow. (a) Control situation: the light entering naturally, (b) the light is deflected by about 90° through mirrors. The bird's direction changed correspondingly. After KRAMER (1950)

If mirrors attached to the window deflected the direction of the sunlight, by approximately 90° (Fig. 43b) the starling also changed its direction by approximately 90°. In training experiments, the bird, again confined in a cylindrical cage excluding a view of landmarks, had to walk to the periphery in a compass direction selected by the experimenter and to look for food invisibly hidden in a special hopper. This method of directional training made experimenting independent of migratory restlessness and its relatively short period of occurrence during the migration season. It has been variously modified and refined. Modifications of the original KRAMER method could also be used with birds lacking migratory restlessness such as homing pigeons. It also permitted, by further experimental interference such as clock shifts, the elucidation of how the sun and the animals' time sense (also called "internal clock") operate as a compass.

HOFFMANN (1954) first demonstrated that the directional response of a trained starling deviated about 90° clockwise (or counter-clockwise) after the bird's clock had been reset by 6h counter-clockwise (or clockwise). Clock shifts are easily accomplished by confining the birds in a closed room under artificial light regimes 6h behind (or 6h ahead of) local time. Corresponding results

56

with pigeons were reported by SCHMIDT-KOENIG (1958, 1961) including an extension to other degrees of clock shifts (Fig. 44), and also including experiments on the time requirement of shifting. A 6-h shift required 4 days and a 12-h shift required 6 days to take full effect.

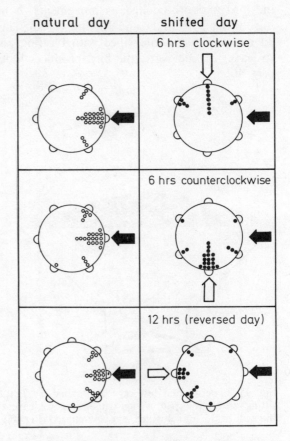

Fig. 44. The effect of clock shifts on the directional response of pigeons in food-rewarded training experiments. In the sketch of the circular training cage with 6 or 8 choice points at the periphery, each open point symbolizes one peck of the pigeon at a peripheral disc under control conditions (natural day), each black dot symbolizes one peck of the pigeon at a peripheral disc under experimental conditions of clock shift as indicated. White arrows indicate the direction expected under experimental conditions, black arrows indicate the training direction. After SCHMIDT-KOENIG (1960)

These and other experiments demonstrated that birds derive compass directions from the sun by allowing for its apparent diurnal movement in azimuth. Altitude and the direction of movement i.e. up or down, from left to right (northern hemisphere) or from right to left (southern hemisphere) were ignored. However, the degree of accuracy of the sun compass and many other questions have not yet been finally answered.

Migratory restlessness was mostly used to demonstrate the bird's ability to use stars for directional reference at night. The first relevant observations were made by direct observation of caged migratorily restless old world warblers *(Sylvia spec.)* (KRAMER, 1949; SAUER and SAUER, 1955). The observer placed himself underneath the circular test cage standing on a tripod and watched the bird and protocolled its movements. A more quantitative approach was introduced by EMLEN and EMLEN (1966). The experimental bird is sitting on an ink pad in a funnel lined with blotting paper (Fig. 45). When attempting to leave the enclosure, the bird produces foot prints which can be evaluated (Fig. 46).

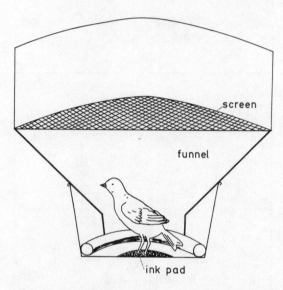

Fig. 45. Experimental test cage as devised by EMLEN and EMLEN (1966). The bird is sitting on an ink pad. When attempting to leave the cage it produces foot prints on the wall of the funnel which can be analyzed (Fig. 46)

In experiments with indigo buntings *(Passerina cyanea)* under planetarium skies EMLEN (e.g. 1972) found that the young bird needs to watch the starry sky and to find out which part rotates least. On the northern hemisphere this is the region around the Pole Star. This part of the sky, perhaps a circumpolar constellation, is, then, used as directional reference. In a crucial experiment EMLEN exposed young buntings to a planetarium sky rotating around a star of Orion (a southern star pattern) rather than around the Pole Star. In the ensuing fall season, these birds headed away from Orion, i.e. into a northerly rather than a southerly direction. In contrast to the sun compass the star compass does not require a clock; clock shifts are ineffective.

Spontaneous migratory restlessness was again used by MERKEL and FROMME (1958); FROMME (1961) and finally by WILTSCHKO (e.g. WILTSCHKO, 1968; WILTSCHKO and WILTSCHKO, 1972) to investigate the orientation of European

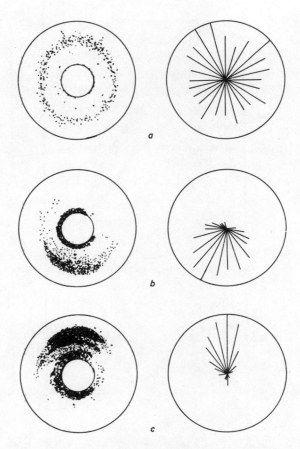

Fig. 46a–c. Three examples of foot print raw data (left) obtained by the funnel technique (Fig. 45) and their translation into vector diagrams (right); (a) little or no direction, (b) rather well-directed orientation to the SSW, (c) well-directed orientation towards NNW. After EMLEN and EMLEN (1966)

robins *(Erithacus rubecula)* without visual cues. An octogonal test cage with radial perches electromechanically recording every hop of the experimental bird was used (Fig. 47). It took several years to establish satisfactorily that some degree of orientation may, indeed, be established or maintained nonvisually. In crucial experiments demonstrating the use of magnetic fields for this capacity, WILTSCHKO was able to alter the directional orientation of the robins predictably by manipulating an artificial magnetic field produced by Helmholtz coils (Fig. 48). WILTSCHKO and WILTSCHKO also presented evidence that the robin's magnetic compass—unlike the human magnetic compass—is based on evaluating the inclination of the axial direction of the field lines to the vertical. The smaller angle between field lines and the vertical is identified as poleward. The polarity of the field is not involved (Fig. 49). A much more detailed review of the compass systems so far investigated has been published by WILTSCHKO (1973).

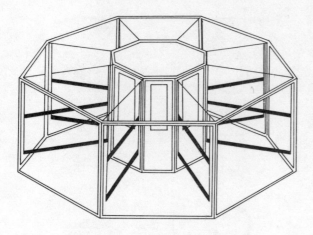

Fig. 47. The octogonal test cage with 8 radial perches used by WILTSCHKO in experiments on the magnetic orientation of European robins. After WILTSCHKO (1968)

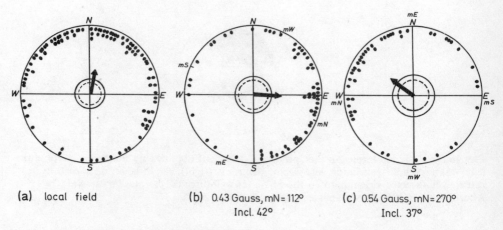

(a) local field

(b) 0.43 Gauss, mN = 112°
Incl. 42°

(c) 0.54 Gauss, mN = 270°
Incl. 37°

Fig. 48a–c. Nonvisual directional orientation of European robins *(Erithacus rubecula)* in spring: (a) local magnetic field unchanged; (b) magnetic N (mN) pointing ESE; (c) magnetic N pointing W. Each dot symbolizes the mean direction of one bird in one night. The arrow originating from the center is the mean vector of all dots with the 1% level (dashed circle) and the 5% level (solid inner circle) for randomness indicated. After WILTSCHKO (1973)

Even though sun, star or magnetic compasses are demonstrated in laboratory experiments, their actual field use cannot be taken for granted without specific proof. It is, however, much easier to experiment with birds in the laboratory than during free flight, actual migration, or homing.

The use of the sun compass in actual homing in pigeons was demonstrated by SCHMIDT-KOENIG (1960, 1961, 1972). The successful laboratory experiments (clock shifts discussed on p. 57) were extended to homing experiments. Upon

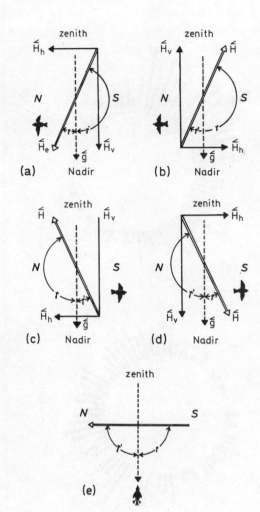

Fig. 49a–e. How a bird may use the inclination of the geomagnetic field for compass orientation (a) local condition, (b) magnetic field with altered polarity; (c) and (d) magnetic field with horizontal component reversed; (e) magnetic field with horizontal field lines. $\hat{H}_e$ = vector of the local magnetic field; $\hat{H}$ = vector of the experimental magnetic field; $\hat{H}_h$ = horizontal component; $\hat{H}_v$ = vertical component; $\hat{g}$ = the vertical; γ = the angle between vertical and magnetic field vector, γ' = supplement angle to γ. From WILTSCHKO and WILTSCHKO (1972)

release, shifted birds deviated in approximately the same fashion from the direction chosen by control birds, as did the trained responses upon shifting from the training direction in the laboratory experiments. Fig. 50 presents a summary of releases involving clock shifts of 6 h clockwise, 6 h counter-clockwise and of 12 h. The effect of a 6-h shift was further tested and confirmed at distances from 200 km to somewhat less than 2 km from the loft (SCHMIDT-KOENIG, 1965; 1969; 1972) and correspondingly less deviation was recorded with shifts of less than 6 h, i. e. shifts of 2 h and of 30 min (SCHMIDT-KOENIG, 1969; 1972). These findings have been confirmed repeatedly by others (e. g., GRAUE, 1963; KEETON, 1969; WALCOTT, 1972), and clock-shifting was for many years the only experimental tool with which it was possible to obtain reliable and reproduceable results on pigeon homing and still remains a powerful experimental method. All results demonstrate overwhelmingly that the sun, if available, is used for compass orientation. If unavailable under overcast skies or at night, other mechanisms must take over.

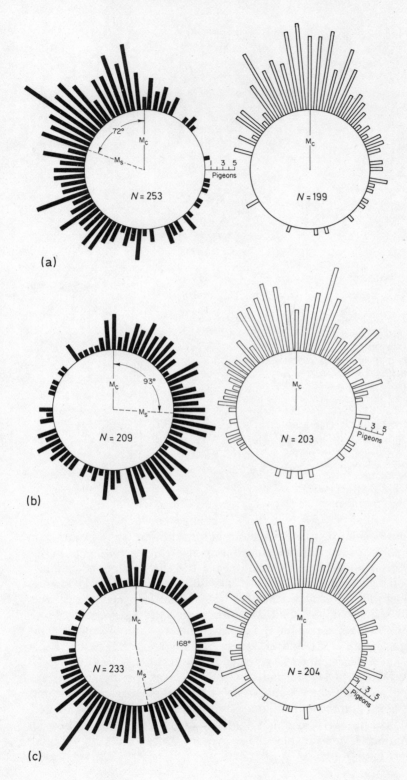

The use of magnetic cues in homing—probably for compass purposes—has been reported by KEETON (1971, 1972). Young pigeons displaced from home for the first time were unable to orient if small magnets were attached to them.

The use of the star compass has been demonstrated to date exclusively in caged birds, not in actual migration. Even the best compass could not be sufficient for navigation, as compasses indicate directions but not goals. Compasses may, however, be integral components of a navigation system. The main component of the system would be the navigational component, i.e., by which the animal establishes its position and the relation between its present position and the goal. Once the direction to the goal has been determined, a compass may be used to reach the goal. The navigational process at the heart of the problem, thus remains unsolved.

If an animal shows some outstanding accomplishment, one must seek some correspondingly well-developed sense organ in order to find the sensory modality involved and to explain its performance and accomplishments. Avian navigation is just such an accomplishment and the avian eye such a well-developed sense organ. For many years, therefore, the eye as the dominating avian sense organ, and vision, respectively, have been associated with or held responsible for the navigational abilities of birds. Experiments directly interfering with vision with flying and navigating birds were, however, not carried out until SCHLICHTE and SCHMIDT-KOENIG (1971), SCHMIDT-KOENIG and SCHLICHTE (1972) succeeded in fitting frosted lenses to the eyes of pigeons thereby reducing vision to near blindness. With lenses in place, the birds were unable to recognize familiar environmental structures at 6 m or more (SCHLICHTE, 1973). Among several series of successful homing experiments those in which the birds were radio-tracked yielded particularly clear results. Birds with frosted lenses were able—although almost blind—to head for home and to reach the vicinity of the loft without much difficulty. Fig. 51 illustrates one example; from SCHMIDT-KOENIG and WALCOTT (1975). At short range however, the birds were unable to pinpoint the loft easily. The bird (W 65) illustrated in Fig. 51 failed to find the loft despite intensive circling while a considerable number of others (not illustrated) finally managed to hit the goal usually after several unsuccessful attempts (SCHMIDT-KOENIG and SCHLICHTE, 1972; SCHMIDT-KOENIG and WALCOTT, 1973). These results provide several important contributions to answering the question how a pigeon may home.

Pigeons seem to use two different systems. (1) A navigation system enables the birds to head for home from distant release sites. This system is surprisingly accurate, guiding the bird to within about 2 km from its goal. It is not noticeably

---

Fig. 50a–c. Summary of initial orientation (vanishing points) and homing performance of homing pigeons in the clock resetting (clock shifting) experiments. Experimentals are given in solid bars, controls in open bars. The time shift involved in series (a), (b), (c) is indicated in the text. The length of bars is proportional to the number of birds that vanished at the bearing indicated. Vanishing bearings of experimentals (Ms for mean of shifted birds) are plotted with reference to the mean of the control (Mc). From SCHMIDT-KOENIG (1960)

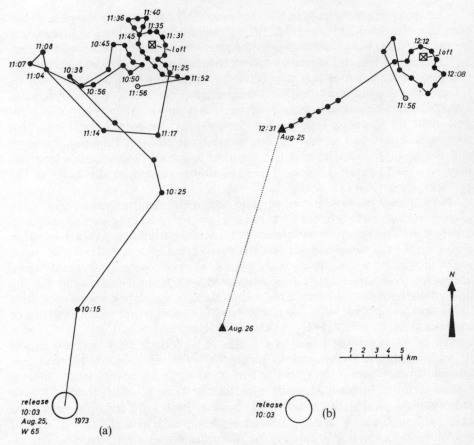

Fig. 51a and b. Track of pigeon W65 wearing frosted lenses released 16 km SSE of its loft at Lincoln, Mass. under sunny conditions. Circular symbols indicate approximate positions of the flying bird at the time indicated, Triangles indicate locations where the bird had perched. Tracking fixes were so numerous that they had to be separated into those until 11:56 (a) and those after 11:56 (b). From SCHMIDT-KOENIG and WALCOTT (1975)

affected by frosted lenses and, hence largely if not entirely nonvisual. The use of familiar landmarks or anything else requiring precise vision can certainly be dismissed. (2) A different system takes over for the final approach to pinpoint the loft. This final approach appears to be severely impaired by frosted lenses and, therefore, to be a largely visual process. A considerable number of birds, however, managed to hit the loft partly by repeated attempts. Thus, even the final approach is not completely obliterated by severe reduction of visual capacities.

Numerous suggestions, theories and complete hypotheses have been advanced to explain how birds navigate. For many years the hypothesis of sun navigation has played an important role in theoretical considerations and experiments. According to the hypothesis advanced by MATTHEWS (1953, 1955, 1968), the displaced bird would extrapolate a short portion of the sun's path to the noon

64

altitude and compare it with that remembered from home. If the noon altitude observed were lower than that at home the bird would be north of home. If the noon altitude were higher than home the bird would be south of home. Similarly, an angular or time difference between actual azimuth position and noon azimuth of the sun as compared to that at home would indicate an east or west displacement. From the numerous experimental results which do not agree with sun navigation two should be mentioned briefly.

(a) The clock-shifting experiments discussed on p. 60–62 unequivocally demonstrate that the sun is used for compass orientation and not for navigation. If the sun were used for navigation, entirely different deviations than those observed would have resulted (SCHMIDT-KOENIG, 1960; 1965; 1969; 1972).

(b) The so-called sun occlusion experiment was introduced by MATTHEWS (1953) in support of this hypothesis: experimental birds were confined in a pen excluding direct view of the sun for 6–9 days at the time of the fall equinox. During this occlusion time, the sun's noon altitude decreased so much that a north displacement was simulated even from a release point not too far south of the loft. The birds should, and most of MATTHEWS' birds did, head south, away from home, in accordance with the hypothesis. This experiment has been repeated by several others (RAWSON and RAWSON, 1955; KRAMER, 1955; 1957; HOFFMANN, 1958; KEETON, 1970) all reporting clearly negative results. Thus, there is no experimental support for MATTHEWS' sun-navigation hypothesis. Moreover, there are arguments that even theoretically MATTHEWS' hypothesis is untenable. The hypothesis requires the bird to measure changes of sun azimuth. It is at present inconceivable how this can be done by a flying bird lacking stable references, for example over the open ocean. But even over land using landmarks, parallactic errors encountered during flight would exceed the quantities to be measured. From the much more detailed discussions of the sun-navigation hypothesis (e. g. KRAMER, 1957; SCHMIDT-KOENIG, 1965, 1970b) the most recent one has been published by KEETON (1974). These repeated discussions are prompted by MATTHEWS' untiring attempts to propagate his hypothesis (MATTHEWS, 1968, 1971).

Another hypothesis suggested the use of inertial forces (BARLOW, 1963). According to this hypothesis the bird keeps track of all outward movements (recording angular acceleration and angular speed) in order either to retrace precisely the outward journey step by step back home or to calculate the direct course home. Although inertial navigation seems not very likely for several reasons not to be discussed here, a few attempts to support or disprove this hypothesis experimentally have been published (e. g. dissection of the horizontal semicircular canal supposedly measuring horizontal acceleration, WALLRAFF, 1965; displacement under full anesthesia, WALCOTT and SCHMIDT-KOENIG, 1973). All had negative results in as much as the experimental birds homed. But the birds may have used other means than the possibly eliminated inertial system. Thus, the finally conclusive experiment has yet to be carried out.

Most recently, PAPI and coworkers (1972) have advanced the hypothesis of homing by olfactory cues. They suggest that pigeons learn, at home, to associate specific odors carried by winds with wind directions. In order to find the home direction upon displacement they would only have to recognize some odor and to reverse the direction from which that odor used to come at home. Well-designed experiments by PAPI et al. (1971; 1972; 1973; 1974), BENVENUTI et al. (1973a, b), BALDACCINI et al. (1974) seem to support the hypothesis. Among a number of aspects apparently incompatible with olfactory navigation (cf. KEETON, 1974) one especially stringent one may be mentioned: the apparent lack of a correspondingly well-developed olfactory system in pigeons (BANG and COBB, 1968). Furthermore, attempts to confirm some of PAPI's results elsewhere (KEETON personal communication; SCHMIDT-KOENIG) have not yet been successful. In view of the early state of this new idea and the relatively small samples of PAPI's, however well designed, experiments it would be premature to try to evaluate the hypothesis of olfactory homing at the present time.

Most hypotheses and very many experiments to verify them have relied primarily on the homing pigeon as a fairly convenient experimental species and possibly a reasonable model for wild birds. Whether or not pigeons are, indeed, realistic models for wild birds has yet to be demonstrated, while their undisputed ability to navigate is by itself certainly a biological problem worth investigation and solution.

There is at present only one idea supported by experimental evidence that is able to explain one particular aspect of migration: how young birds may reach the winter range on their first fall migration. PERDECK's experiments with starlings discussed on p. 55 indicated the prevalent role of directional orientation. If, in addition, a bird has some information on the distance to be flown, vector navigation or vector orientation, a limited kind of navigation, would be possible (it is a matter of definition whether or not it deserves the name navigation). In a series of laboratory experiments with migratorily restless old world warblers (genus *Phylloscopus* and *Sylvia*) GWINNER (1968a, b; 1972) found a relation between migratory restlessness (as time or activity) expended in the recording cage and the migratory distance to be covered out of doors. Willow warblers *(Phylloscopus trochilus)* wintering in southern Africa spent correspondingly more time or activity than wood warblers *(Phylloscopus sibilatrix)* wintering in central Africa and those spent correspondingly more than chiffchaffs *(Phylloscopus colybita)* wintering in northern Africa. There is more and more evidence that the seasonal activities of birds are controlled by a circannual clock (analogous to a circadian clock) but details are as yet obscure, and it is entirely unknown how the birds measure activity, energy or distance. The accuracy of such a system certainly cannot be expected to be overwhelming, but using information on the direction plus information on the distance such a system, may, in principle, operate successfully.

More detailed reviews on bird orientation and navigation than the present contribution have been published by SCHMIDT-KOENIG (1965, 1973), GRIFFIN (1969), KEETON (1974) and EMLEN (1974).

Although a large proportion of all theoretical and experimental attempts to elucidate animal navigation has been devoted to birds we still lack a satisfactory insight into the problem even with birds and much work still remains to be done.

9. Mammals

I. Bats

A. Field Performance in Orientation

Although bats are flying animals, their migratory performance is much below that of long-distance migrants such as some marine mammals, birds, and fishes. Nevertheless, some bats do migrate seasonally and a number of displacement

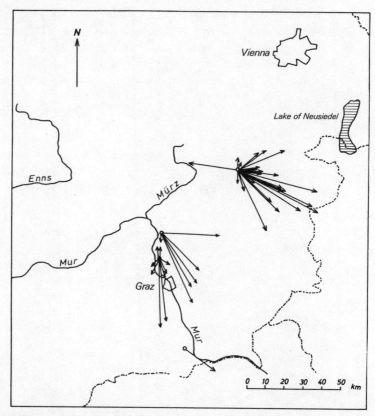

Fig. 52. Recoveries of *Rhinolophus hipposideros* banded in their wintering caves. After KEPKA (1960) and MRKOS (1962)

experiments have demonstrated that at least some bats are good homers and, hence, have their place in a book on migration and homing in animals.

Bats of the temperate zone may be divided into two groups. The members of one group move seasonally but stay within the same climatic zone, moving at best to and from suitable locations for hibernation (caves, buildings). *Rhinolophus hipposideros* banded in Austria may serve as an example of this kind of migration. The longest distance recorded hardly exceeds 40 km (Fig. 52). There seems to be a rather pronounced drive to the south and southeast for the spring movement. Owing to the nocturnal and secretive life of bats, many questions, even those on migratory movements have yet to be answered. Members of the other group migrate to regions where the climate enables them either to hibernate under conditions more favorable than they would encounter when staying all year in their summer range, or even to remain active throughout the winter. Bats of this type are found in areas such as North America, where zones with sufficiently warm winters can be reached without major geographical barriers such as mountain ranges or large bodies of water to be crossed.

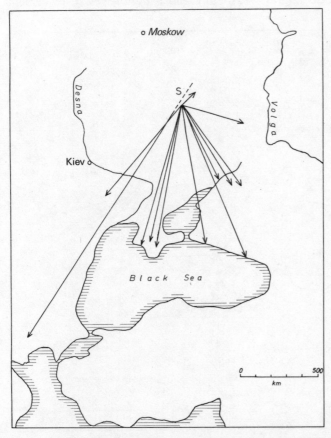

Fig. 53. Recoveries of *Nyctalus noctula* banded in their summer residential area. From HEDIGER (1967), after FORMOSOV, BURESCH and BERON

68

Red bats *(Lasiurus borealis)*, some of their relatives, and some other species inhabiting the North American continent are considered to migrate over fairly long distances. However, not the entire population seems to leave; some portion may hibernate at home. Detailed evidence from banding seems to be scarce and the information available lags much behind that on bird migration. Somewhat more information from banding is available on the European *Nyctalus noctula*. Some populations of this bat, which is a rather powerful flier, seem to migrate fairly long distances from summer territories with rather harsh continental winter climate to areas with mild winter temperatures (Fig. 53).

Despite this incomplete picture of migration in bats, considerable evidence has been accumulated on homing to roosts.

B. Experimental and Theoretical Analysis

The bulk of research on bats has been concerned with the analysis of their sonar systems. Bats with sonar capability use it to avoid obstacles and to locate prey, mostly nocturnal insects. Except for some overlaps, research on homing is a different field of investigation.

Although our knowledge on bat migration is unsatisfactorily sparse, experiments to explain migration and, particularly, the regular homing to roosts have been rather numerous. Several authors have displaced various species of bats from their roosts and have recorded their returns (cf. COCKRUM, 1956; GIFFORD and GRIFFIN, 1960; WILLIAMS et al., 1966; MUELLER, 1966). Homing speeds were rather low and did not convincingly exclude the possibility of more or less random search being the prevalent component rather than directed orientation. More refined methods such as radio tracking have been employed by WILLIAMS and WILLIAMS (1967). Small transmitters were attached to *Phyllostomus hastatus* which easily carried these packages of 7 g. A sample of approximate tracks is given in Fig. 54. Although other reported tracks did not point homeward quite as clearly as the examples given in Fig. 54, at least *Phyllostomus hastatus* appears to home over relatively short distances by directed orientation rather than by random search. Homing success (i. e. number of animals home from total number released) dropped as a function of the distance of displacement to about 10 % at 65 km distance.

Additional experimental interference effected by various authors, such as blindfolding, plugging of ears or noses and using different species of bats has not been too successful in explaining how bats navigate. At least some blindfolded bats were able to home (MUELLER, 1966). Only interference with the sonar system resulted in clearcut results: all experimental bats refused to fly. This does not mean that sonar is involved in homing; it may just mean that the bats refuse to fly when they are deprived of their customary means of avoiding obstacles. No evidence is available on the range of the various sonar systems developed among bats i. e. on the distance at which a bat may identify some familiar landmark. Likewise, information how far away bats "know" landmarks, i.e. whether or not they have some sort of a map of familiar landmarks, is lacking.

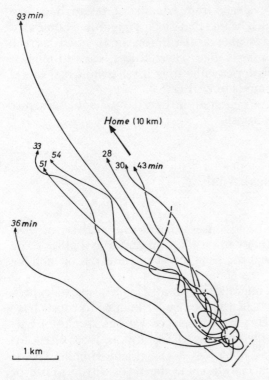

Fig. 54. Sample flight paths of *Phyllostomus hastatus* released 10 km from their roost on Trinidad recovered by radio tracking. The tracks are approximated from a tracking accuracy of ±10°. The numbers at the end of each track give the time elapsed between release and signal lost (arrow head). Tracks dashed at their distant end indicate that the signal was lost prematurely. From WILLIAMS and WILLIAMS (1967)

II. Whales

A. Field Performance in Orientation

Among long-distance migrants, mammals hold a considerable place. Naturally, marine mammals such as whales, seals and the like, inhabiting the most suitable medium for easy transportation of heavy weights, lead the list. Among these, some whales compare well with birds in long-distance migration. A well-known and fairly well-studied migratory species of whales is the grey whale *(Eschrichtius*

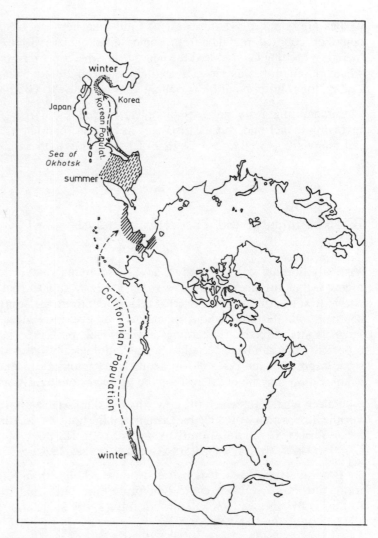

Fig. 55. Summer ranges and winter breeding ranges of the Korean and the Californian populations of the grey whale *(Eschrichtius gibbosus)*. From ORR (1970), after GILMORE

gibbosus). Two populations are known to inhabit parts of the northern Pacific (Fig. 55). The Korean population has its summer feeding range in the Sea of Okhotsk. In fall, the population migrates south to its breeding grounds in warm waters suitable for the birth of calves lacking the insulating blubber essential for survival in the cold waters of the summer feeding area. When the calves have grown somewhat and have accumulated sufficient insulation, the population migrates north again in spring.

A very similar pattern is evident for the Californian population having its summer range in the Bering Sea and its breeding grounds in the shallow warm waters at the California peninsula (Fig. 55). Around 1850 about 1,000 grey whales migrating along the coast of California between Dec. 15, and Feb. 1, could be counted per day from some suitable observation points. By 1875, merciless hunting had reduced the figure to 40 whales per day and the Californian whale population was close to extinction when hunting was finally banned. Today, the total population is estimated at close to 20,000 animals.

Similar migratory patterns of other smaller and larger marine mammals involving longer and shorter distances have been reported in the literature but not many are as well known as that of *Eschrichtius gibbosus*.

B. Experimental and Theoretical Analysis

Virtually nothing is known as to how these animals navigate. To explain the migration of the grey whale one may simply postulate that they follow the coast at some distance or over a certain depth range, north in spring and south in fall. But what about other more or even exclusively pelagic species? Experiments directly concerning navigational problems have hardly been reported. They can be expected to be methodologically rather difficult. However, some aspects of the research on acoustic communication and on short-range echolocation may be of interest for the problems of navigation.

Baleen whales (of which the grey whale is one) seem to have a low-frequency vocalization which would enable them to tell the bottom, the surface, and possible large objects by echolocation (REYSENBACH DE HAAN, 1966; WATKINS, 1966; CUMMINGS et al., 1968; CUMMINGS and PHILIPPI, 1970).

Porpoises have been demonstrated to use echolocation (KELLOG, 1961), and toothed whales are also likely to echolocate, since they emit sounds very similar to those of porpoises (BACKUS and SCHEVILL, 1966; BULLOCK and RIDGWAY, 1972; NORRIS and HARVEY, 1972).

Whether or not pinnipeds (seals) also use echolocation has not yet been firmly established. In any case, echolocation, especially in the more refined form employed by porpoises and toothed whales, may be used for recognition of underwater landmarks i.e. for some kind of piloting; but this has, so far, not been demonstrated. The echolocation of the sperm whale is especially considered to be a long-range echolocation system possibly penetrating a few hundred meters, primarily used for food location in the deep sea. Other evidence, on which at least some speculation could be hinged as to how marine mammals find their ways, does not seem to be available.

III. Terrestrial Mammals

A. Field Performance in Orientation

Most readers will have heard of the mass movements of lemmings (*Lemmus* spec.), arctic rodents related to the voles, usually in three- to four-year cycles, overpopulation apparently inducing mass emigrations. These are sometimes spectacular, especially when large numbers move on into rivers and lakes and drown as if a distant goal attracted them with such force that they ignore all perils of such natural obstacles. These eruptions are, however, irregular and they end in destruction without return; their guiding factors are unknown. While small mammals have been preferably used in laboratory experiments (p. 74) it is much more rewarding to turn to larger mammals for information on migratory movements.

Tagging has confirmed observations that many large terrestrial mammals cover considerable distances annually. The migrations of these exclusively herbivorous animals are, as a rule, related to the availability of food and/or water in different locations which is, in turn, mostly a seasonal affair. As examples of large movements of several hundred to over a thousand kilometers, the migration of zebras *(Equus quagga)* and of gnus (*Connochaetes* spec.) in the Serengeti, as investigated by GRZIMEK and GRZIMEK (1960), or the somewhat less extensive migrations of caribous *(Rangifer tarandus)* in parts of Alaska and Canada (e. g. KELSALL, 1968) may be mentioned. Movements involving less horizontal but more vertical distances i. e. up into the mountains in spring and down again in fall are common in mountainous regions. Migrations of this type have been reported in some detail, for example by EDWARDS and RITCEY (1956) for Canadian moose *(Alces alces)*. The maximum distance between the higher summer grounds and lower wintering grounds is about 65 km, and according to these authors equivalent to 550 km of horizontal travel. Elk *(Cervus elaphus canadensis)* of the Jackson Hole Refuge have similar records. One striking feature in these migrations is that the animals march in long lines. Detailed data of the movement of individual animals are now being collected by radio tracking. This leads to the chapter on experimental and theoretical analysis.

B. Experimental and Theoretical Analysis

Large mammals certainly invite the investigator to attach a transmitter and to record their movements by radio tracking. In early states of this art nonmigratory animals such as grizzly bears *(Ursus arctos horribilis)* (CRAIGHEAD et al., 1963; CRAIGHEAD and CRAIGHEAD, 1965) or woodchucks *(Marmota monax rufescens)* (MERRIAM, 1963), were tracked over relatively short distances by ground-based or man-carried antennas. The limitations of short-range tracking prevented tracking of long-distance migrants over longer distances. Long-distance tracking is now being accomplished via satellites, the first fairly successful attempts involving an elk cow *(Cervus elaphus canadensis)* in the Jackson Hole Refuge in Wyoming, USA and satellites of the NIMBUS series (CRAIGHEAD et al., 1972). Thanks to miniaturization in the wake of space flights, the size and weight of transmitter packages was considerably reduced. Large animals may now be instrumented more easily and smaller animals may be included in the list of animals which can be radio-tracked in long-distance, long-range and long-time projects. Some transmitters will also provide environmental and physiological information such as ambient temperature, heart rate, body temperature and the like. Despite these very promising developments in techniques and instrumentation nothing is known to date on how elk, moose, caribou, zebra, gnus and others find their way.

Understandably, large mammals such as those discussed above lend themselves less readily to laboratory orientation experiments than smaller mammals such as mice, cats or dogs. Newspapers and magazines periodically report on homing of dogs and cats to their old home after having been moved to some new home sometimes several hundred kilometers away. One would imagine that dogs and cats had been used extensively in homing experiments, but surprisingly this is not so, and hard facts to check the homing ability of these pets seem to be non-existent. Cats have been used in maze experiments by PRECHT and LINDENLAUB (1954). Leaving the maze in the home direction was just slightly better than chance at 5 km from home and no better than chance at 12 km. Since visual and olfactory cues were excluded, these experiments are not entirely relevant to discount the newspaper reports. True homing experiments are needed.

Mice have been used in laboratory experiments rather extensively although their record of homing ability in the field is poor. If displaced and released, only a negligible proportion of several investigated species such as the meadow mouse *(Microtus pennsylvanicus)*, the western harvest mouse *(Reithrodontomys megalotis)*, and the deer mouse *(Peromyscus maniculatus)* homed over a few hundred meters according to ROBINSON and FALLS (1965); FISLER (1966, 1967), and BOVET (1968, 1971). These results do not convincingly exclude the possibility that successful homing over several hundred meters was accomplished by chance. But LINDENLAUB (1955, 1960) and BOVET (1960, 1968 and 1971) present some evidence that initial orientation upon release in the field or in a radially symmetrical and optically screened maze bears some, however weak, relation to the home direction. Such a relation would indicate the operation of some sort of non-random orientation mechanism. The nature of this mechanism and the cues used by the mice remained, however, entirely obscure.

After the sun compass had been discovered first in bees (V. FRISCH, 1950) and birds (KRAMER, 1950) almost every other animal of interest for orientation studies and useful for laboratory experiments has been tested for possible use of the sun compass. Mammals seem to present an exception. Only the striped field mouse *(Apodemus agrarius)* which is also to some degree diurnally active has been subject to modestly successful sun-compass experiments (LÜTERS and BIRUKOW, 1963). Other nocturnal mammals such as bats or clearly diurnal mammals for which a sun compass would be useful have, apparently, not been investigated. While we have evidence in most relevant animals that they dispose of one or more compasses, in most mammals even this information is lacking.

Conclusion

The migrations of animals are among the most fascinating of environmental phenomena. In the interest of conservation it is becoming increasingly important to have a detailed understanding of the seasonal (or aseasonal) movements of animals. Nevertheless, even this purely descriptive information is as yet unsatisfactory and incomplete for many species. An important aspect of animal migration is that of orientation and navigation, the understanding of which requires complementary experimental investigations. Even for those species for which the routes and schedules are well known (e.g., some species of birds, fish, and turtles) information on the mechanisms of navigation is lacking. The history of research on orientation includes a number of apparent solutions to the problems of navigation that have subsequently proved to be only pseudo-solutions. Examples include the hypotheses of bird navigation of YEAGLEY (1947, 1951), of MATTHEWS (1953, 1955, 1968) and of SAUER (1957, 1961, 1963; SAUER and SAUER, 1960). YEAGLEY suggests the use of components of the earth's magnetic field and Coriolis force, MATTHEWS the use of variables in position of the sun (p. 64, 65), and SAUER variables in the positions of stars as information in navigation by birds. All three investigators presented experimental evidence that apparently supported their hypotheses, but repetition of the experiments by other investigators produced contrary results. Research has revealed only a few modes of orientation in relatively complete detail. Homing in bees, as elucidated by v. FRISCH (1965, 1968), his students and their students, is the master example. For other species, partial solutions of the problems of orientation have been found; included are the use of a solar, stellar or magnetic compass for directional orientation; olfactory homing in some species of fish; or wind-drift in locusts. Exciting advances have been made recently in this field of research. But there still remains a plethora of as yet unanswered questions; questions that are very basic to the problems of animal navigation and homing are still unanswered. We can hope that advances in modern technology, new ideas and last, but not least, continuing support will eventually help to overcome the sometimes formidable methodological barriers arising from the distances involved and from the mobility of the animals—to the heart of the problem of how animals navigate.

Appendix: Some Statistical Methods for the Analysis of Animal Orientation Data

Most observations on animal orientation, either from experimental or natural aspects, are plagued with a high degree of variability. Moreover, many observations are obtained as directional (i. e., two-dimensional or circular) data, a special field that attracted the interest of statisticians rather late. For many years, relevant publications were scattered in the literature and important problems remained uninvestigated. Useful summaries of statistical methods for directional data have appeared only recently (BATSCHELET, 1965, 1972; MARDIA, 1972). I shall review here some important and frequently used statistical methods as an outline of today's possibilities in dealing with directional data.

It should be emphasized that experiments discussed in the body of this book are of vastly different vintage and quality. In many of them, appropriate statistical methods were not and cannot even be applied. Even diagrams and figures may be unsatisfactory or not as good as they could be on the basis of contemporary standards.

I. A One-Sample Test in Unimodal Distributions

Frequently, an animal's path is tracked in the field in some way (e. g. a turtle in the ocean, an elk on its seasonal migration, or a pigeon on its way home) and the question arises whether the movement is directed or random. Fig. 56a presents an example of a recorded homing flight of a pigeon (SCHMIDT-KOENIG and WALCOTT, 1975). The bird did get home, its movements were certainly oriented, but the data might just as well be from a butterfly with unknown destination and contact was lost at the point where the pigeon happened to be at home.

Data like those of Fig. 56a lack one essential requirement for statistical analysis: independence among data, i. e. among consecutive sections of the animal's track. The animal does not change headings at random, but each previous heading influences the following one. Although such facts cannot be effectively overcome, application of a test by HODGES and AJNE (cf. BATSCHELET, 1972) permits an approach that may be acceptable despite some reservations. The null hypothesis states that the headings were taken from a uniform distribution, i. e. that they have random directions.

As demonstrated in Fig. 56 (b), the sections of the animal's track of Fig. 56 (a) may be plotted as vectors on polar coordinates. A line (dashed in Fig. 56 (b)) is then drawn through the origin and rotated until the number of vectors

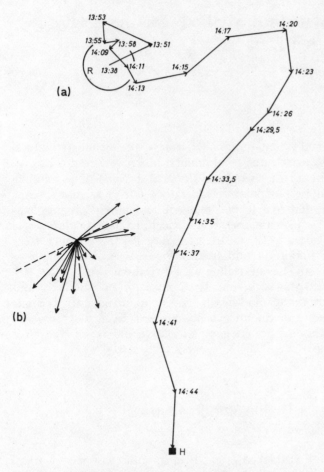

(a)

(b)

Fig. 56. (a) Track of a pigeon released at R (from SCHMIDT-KOENIG and WALCOTT, 1975) plotted as vectors. The time of day is given at each vector head, the 19 vectors are plotted in (b) for the test by HODGES and AJNE. For more details, see text

on one side of the line reaches the possible minimum. This minimum number of vectors on one side of the line is the test statistic K. In our example $K = 2$, $n = 19$. Fig. 57 presents critical values of K for the 1% and for the 5% level as a function of n. At $n = 19$ and $K = 2$ we obtain $p = 0.01$, we reject the null hypothesis and conclude that the track was nonrandom, i. e. directed.

In the majority of cases, directional data are recorded on a circle as unit vectors α; i. e. with radius $= 1$ and plotted as dots (Fig. 58). If such an empirical, circular distribution of independent data is unimodal, the data may be processed by vector addition. Many experiments are performed in the field with reference to true North $= 0°$, East $= 90°$, South $= 180°$, West $= 270°$. We therefore follow this clockwise pattern rather than the pattern of the mathematicians who start at East $= 0°$ and go counter-clockwise.

80

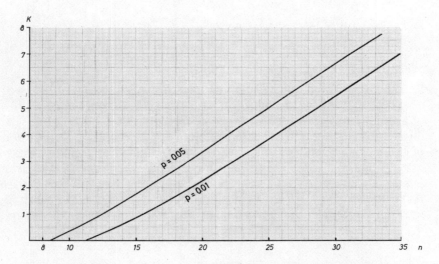

Fig. 57. Critical values of K (ordinate) for 5% and for 1% probability as a function of sample size (abscissa) for the test by HODGES and AJNE. Figures from BATSCHELET (1972)

The mean direction of a circular sample is obtained as

$$\tan g\alpha = \frac{\Sigma \sin \alpha_i}{\Sigma \cos \alpha_i} \qquad (1)$$

The ambiguity of the table value of $\tan g\alpha$ must be solved by inspection of the data and by accounting for the sign of sin and cos in the four quadrants ($\tan g\alpha = 2.02; \alpha = 63°38'$ in Fig. 58).

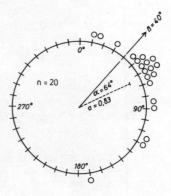

Fig. 58. A unimodal circular sample of 20 scores and the graphical presentation of two statistics obtained through vector analysis: the mean vector with $\alpha = 64°$, and $a = 0.83$; β would be an expected direction

The length a of the mean vector is obtained as

$$a = \frac{R}{n} \qquad (2)$$

81

and the length of the resultant vector

$$R = \sqrt{(\Sigma \sin \alpha_i)^2 + (\Sigma \cos \alpha_i)^2} \tag{3}$$

(in Fig. 58 $R = 16.65$; $a = 0.83$). In the literature a is frequently called r.

The length of the mean vector is a measure of concentration. If all data of a sample were recorded in the same direction, then $a = 1.0$, if the data were distributed in an ideally random fashion, $a = 0$; thus $0.0 \leq a \leq 1.0$. The length of the mean vector a may be used as test statistic of the RAYLEIGH test to prove whether a sample has been drawn from a uniform (H_0) or from some unimodal population (H_1). For references see SCHMIDT-KOENIG (1961) and BATSCHELET (1965; 1972). Critical values for 1% and 5% probability, as a function of sample size, are given in Fig. 59. Accordingly, the sample given

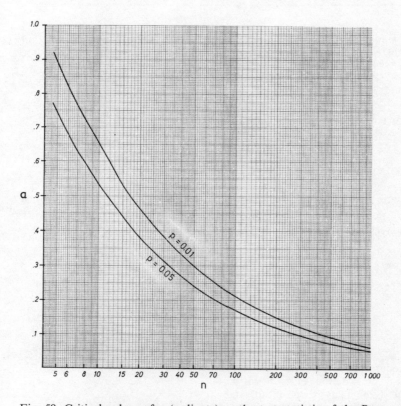

Fig. 59. Critical values of a (ordinate) as the test statistic of the RAYLEIGH test for 5% and for 1% probability as a function of sample size (abscissa; logarithmic scale). From GREENWOOD and DURAND (1955) and SCHMIDT-KOENIG (1961)

in Fig. 58 is highly nonrandom ($p < 0.01$). For the sample given as Fig. 33b (nocturnal release of parrot fish) we find $p > 0.05$ i.e. we have no reason to reject H_0 and assume randomness in the animals' movements.

Often, a specific direction is expected in advance in an experiment. The home direction in a homing experiment (β in Fig. 58); a certain theoretically expected angle to a stimulus such as a light source, gravity or magnetic lines, or a long-standing empirical mean direction are examples of directions that may be considered, preferred, or expected directions. Using this additional information (lacking in the RAYLEIGH test) the V test shows whether an empirical sample is uniformly distributed (H_0) or clustered around the expected direction (H_1). For references see KEETON (1971) and BATSCHELET (1972). If the angle to the expected direction is denoted β (e.g. 40° in Fig. 58 or 70° in Fig. 61), V is the cos component of the resultant vector R of the sample,

$$V = R \cdot \cos(\alpha - \beta). \tag{4}$$

For practical purposes it is easier to use

$$u = \sqrt{\frac{2}{n}} \cdot V \tag{5}$$

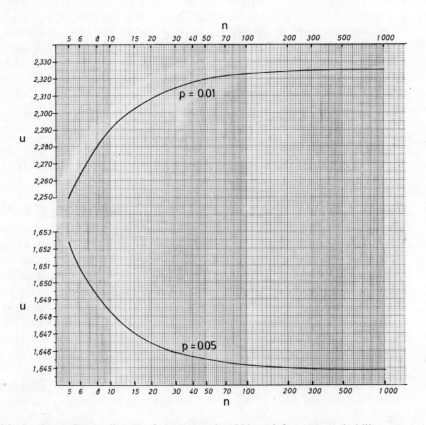

Fig. 60. Critical values of u (ordinate) of the V-test for 5% and for 1% probability as a function of sample size (abscissa; logarithmic scale). Data taken from a Table by W. T. KEETON, published by BATSCHELET (1972)

rather than V as test statistic. Critical values of u for 1% and for 5% probality are given in Fig. 60. In marginal cases this more powerful test may yield significance while the RAYLEIGH test would not. Such an example is given in Fig. 61. Ten

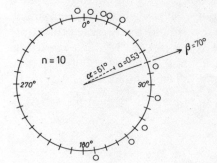

Fig. 61. A circular sample of 10 scores, mean vector $=61°$; $a=0.53$ and an expected direction $\beta=70°$

scores are scattered through about one half of the circle. Before the RAYLEIGH test was available, this sample would have been interpreted as clearly oriented in the eastern half of the circle. With $a=0.53$, however, the RAYLEIGH test at $n=10$ shows $p>0.05$ which means not significantly different from uniformity. But when a theoretical direction is available at 70° and the V test is applied ($V=5.263$; $u=2.353$), Fig. 60 reveals $p<0.01$. Thus, the sample is significantly concentrated in the theoretical direction.

II. A One-Sample Test in Bimodal or Multimodal Distributions

So far we have dealt with situations in which unimodal and uniform distributions were the two alternatives. Frequently, however, bimodal or multimodal distributions occur in orientation experiments as, for example, when animals orient to the plane of polarized light or on a north-south axis without the capacity to discriminate north and south. In such cases we may use RAO's test (RAO, 1969 cf. BATSCHELET, 1972) to find out whether a sample has been drawn from a uniform (H_0) or from a multimodal (H_1) population; first, we calculate the arcs or angles (in degrees) between all consecutive scores (α_i) on the circle and denote these angles by T_i. $T_1=\alpha_2-\alpha_1$; $T_2=\alpha_3-\alpha_2$ and so on. Note that $\Sigma T_i=360°$. If the scores are distributed in a uniform fashion, the T_i should vary more or less closely around their arithmetic mean $\bar{x}=(360°/n)$. If, however, the scores are concentrated in one or more directions, the T_i should deviate from $360°/n$. The sum of these deviations may be used to calculate the test statistic U.

$$U = \frac{1}{2} \Sigma \left[T_i - \frac{360°}{n} \right] \tag{6}$$

84

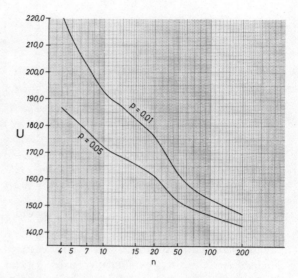

Fig. 62. Critical values of U (ordinate) for RAO's test as a function of sample size (abscissa). Figs. taken from a Table in BATSCHELET (1972)

Critical values of U for 1% and for 5% probability as a function of sample size are given in Fig. 62. Fig. 63 presents an example that appears to be bimodal by visual inspection. Application of the RAO test yields

$$\frac{360°}{33} = 10.91°; \qquad \Sigma\left[T_i - \frac{360°}{n}\right] = 331.83; \qquad U = 165.92;$$

entering Fig. 62 we find $p \simeq 0.01$, i.e. we accept nonrandomness (H_1). If we had calculated the mean vector $(a = 0.08)$, which would be meaningless in any multimodal distribution, the RAYLEIGH test would have indicated $p > 0.05$, i.e. we would have to accept uniformity (H_0).

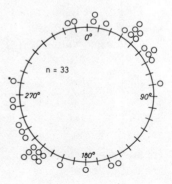

Fig. 63. A bimodal circular sample of 33 scores obtained by WAGNER (1968) with homing pigeons released on Lake Constance half way between the northeastern and the southwestern shores

III. Two-Sample Tests

So far we have discussed so-called one-sample tests by which an empirical sample is tested against a theoretical population. The so-called two-sample tests may be applied to test whether two empirical samples have been drawn from the same (H_0) or from two different (H_1) populations i.e. whether or not two empirical samples are different. This question very frequently arises when experimental and control groups are to be compared.

A powerful test that does not postulate a specific form of distribution (such as circular normal distribution) and requires only that the data be independent of each other is the WATSON test (WATSON, 1961; 1962). It is a so-called omnibus test, sensitive to any kind of difference in direction, scatter or skewness. The calculations required are tedious even if efficient table calculators are available. The test may be more efficiently used for large samples or larger series of samples if programmed. A hand-calculated example (not illustrated as circular diagram) is presented as Table 2. Sample sizes (n and m) do not need to be equal as happens to be the case in this example. The scores of the two samples (α_i and α_j) are arranged in numerical order as shown in Column 1 and 2, then the fractions i/n and j/m are formed (Column 3 and 4) and the number of inter- and intrasample ties is determined (Column 5). From these figures we calculate

$$S_1 = \sum_k \left[\left(\frac{i}{n} - \frac{j}{m} \right) \cdot n_k \right] \quad \text{and} \tag{7}$$

$$S_2 = \sum_k \left[\left(\frac{i}{n} - \frac{j}{m} \right)^2 \cdot n_k \right]. \tag{8}$$

In our example $S_1 = -2.15$; $S_2 = 0.5925$.

Then the test statistic $U_{n,m}^2$ may be calculated:

$$U_{n,m}^2 = \frac{n \cdot m}{(n+m)^2} \cdot \left(S_2 - \frac{S_1^2}{n+m} \right). \tag{9}$$

In our example $U_{20;20}^2 = 0.1192$.

Critical values cannot be presented as a chart. The reader is, therefore, referred to an extensive Table published by BATSCHELET (1972), from which a few selected values are given here for the special case of $n=m$.

$n=m$	10	20	30	40	50	∞
1 %	0.250	0.259	0.262	0.263	0.264	0.2684
5 %	0.184	0.185	0.186	0.186	0.186	0.1869

In our example with $U_{20;20}^2 = 0.1192$ we find $p > 0.05$, hence the two samples are not different.

86

Table 2. Example of the WATSON-test. Vanishing bearings of two groups of pigeons are given in Column 1 and 2 in numerical order; intrasample ties are indicated by $2 \times$. More explanations are given in the text

1	2	3	4	5
		i/n	j/m	n_k
35°		1/20	0/20	1
120°		2/20	0/20	1
130°	130°	3/20	1/20	2
135°		4/20	1/20	1
	155°	4/20	2/20	1
	160°	4/20	3/20	1
	165°	4/20	4/20	1
230°		5/20	4/20	1
	235°	5/20	5/20	1
	250°	5/20	6/20	1
	$2 \times 265°$	5/20	8/20	2
	270°	5/20	9/20	1
	$2 \times 280°$	5/20	11/20	2
$2 \times 285°$		7/20	11/20	2
290°	290°	8/20	12/20	2
295°		9/20	12/20	1
$2 \times 300°$	300°	11/20	13/20	3
305°	305°	12/20	14/20	2
310°	310°	13/20	15/20	2
315°		14/20	15/20	1
320°		15/20	15/20	1
325°	325°	16/20	16/20	2
330°	330°	17/20	17/20	2
335°	335°	18/20	18/20	2
345°	345°	19/20	19/20	2
350°	350°	20/20	20/20	2

$n = 20$

$m = 20$

IV. Multisample Tests

Sometimes more than two samples have been obtained in an experiment and the question has to be answered whether they are different from each other. WATSON and WILLIAMS (1956) have proposed a test that is sensitive to differences in the mean direction of q independent samples. This test is a so-called parametric test, i. e. the samples have to be drawn from circular normal distributions with the same measure of dispersion[1].

[1] How the experimenter can test whether or not a sample has been drawn from a normally distributed circular population may be taken from BATSCHELET (1965).

The test requires calculation of the length of the resultant vector R_i of each sample (Eq. (3)) and combination of all samples and calculation of R_N for the pooled samples (Eq. (3)). The sample size of the combined sample would be $N = \Sigma n_i$. The test statistic is calculated as

$$F = \frac{(N-q)(\Sigma R_i - R_N)}{(q-1)(N-R_i)} \tag{10}$$

Critical values of $F_{q-1;N-q}$ may be taken from Tables of the F distribution to be found in almost any textbook on statistics or collections of statistical tables. As an example Fig. 64 presents 3 vanishing diagrams of pigeons that

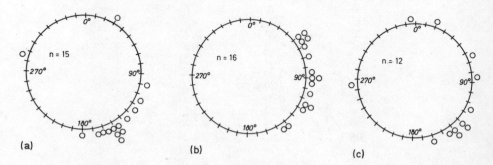

Fig. 64a–c. Three circular samples (vanishing diagrams of initial orientation) obtained during a homing experiment with pigeons homing to three lofts in the same direction but in different distances from the release site. Sample raw data from the "cross loft experiment", SCHMIDT-KOENIG (1963)

had the same direction but different distances to home (sample raw data from the "cross loft experiment" by SCHMIDT-KOENIG, 1963).

The data could as well be taken from similar experiments with mice, salamanders or sandhoppers. In this example $R_1 = 12.85$; $R_2 = 6.37$; $R_3 = 13.83$; $N = 43$; $q = 3$; $R_N = 28.74$; $F_{2;40} = 2.61$. The tabulated values are $F_{0.01} = 5.18$ and $F_{0.05} = 3.23$, thus we have to accept H_0. These three samples are homogenous, i. e. their mean directions do not differ significantly.

These were but a few examples to outline the possibilities nowadays available for the statistical treatment of circular data. For many more and different kinds of statistical methods see BATSCHELET (1965; 1972) and MARDIA (1972), the latter text being more theoretical and mathematical than the former. For the statistical analysis of non-circular data obtained in studies of orientation phenomena such as distances, time intervals, speeds, altitudes and the like the reader may be referred to any of the many texts of statistics, especially those with major emphasis on nonparametric statistics. From these, the χ^2-tests, the Mann-Whitney-U-test, the matched-pairs-signed-ranks test, the Kolmogorov-Smirnov-test, and rank order correlation methods are most frequently used.

References

ALM, G. (1958): Seasonal fluctuations in the catches of salmon from the Baltic. J. Cons. perm. int. Explor. Mer. **23**, 399–433.

BACKUS, R. H., SCHEVILL, W. E. (1966): Physeter clicks. In: Whales, Dolphins and Porpoises. K. S. NORRIS (ed.), Univ. of Calif. Press, 510–528.

BALDACCINI, N. E., BENVENUTI, S., FIASCHI, V., IOALÉ, P., PAPI, F. (1974): Pigeon homing: Effects of manipulation of sensory experience at home site. J. comp. Physiol. **94**, 85–96.

BALMAIN, K. H., SHEARER, W. M. (1956): Records of salmon and sea trout caught at sea. Freshwat. Salm. Fish. Res., **11**, 12–40.

BANG, B. G., COBB, S. (1968): The size of the olfactory bulb in 108 species of birds. Auk **85**, 55–61.

BARLOW, J. S. (1963): Inertial navigation as a basis for animal navigation. J. theoret. Biol. **6**, 76–117.

BATSCHELET, E. (1965): Statistical methods for the analysis of problems in animal orientation and certain biological rhythms. The American Institute of Biological Sciences, Washington, D. C.

BATSCHELET, E. (1972): Recent statistical methods for orientation data. In: Animal Orientation and Navigation. S. R. GALLER, K. SCHMIDT-KOENIG, G. J. JACOBS, R. E. BELLEVILLE (eds.): National Aeronautics and Space Admin. Wash. D. C.: 61–91.

BENVENUTI, S., FIASCHI, V., FIORE, L., PAPI, F. (1973a): Disturbance of homing behavior in pigeons experimentally induced by olfactory stimuli. Monit. Zool. Ital. (N.S.) **7**, 117–128.

BENVENUTI, S., FIASCHI, V., FIORE, L., PAPI, F. (1973b): Homing performances of inexperienced and directionally trained pigeons subjected to olfactory nerve section. J. comp. Physiol. **83**, 81–92.

BIRUKOW, G. (1957): Lichtkompassorientierung beim Wasserläufer *Velia currens F.* (Heteroptera) am Tage und zur Nachtzeit I.: Herbst- und Winterversuche. Z. Tierpsychol. **15**, 265–276.

BIRUKOW, G. (1960): Innate types of chronometry in insect orientation. Cold Spring Harb. Symp. **25**, 403–412.

BIRUKOW, G., BUSCH, E. (1957): Lichtkompaßorientierung beim Wasserläufer *Velia currens F.* am Tage und zur Nachtzeit II.: Orientierungsrhythmik in verschiedenen Lichtbedingungen. Z. Tierpsychol. **14**, 184–203.

BIRUKOW, G., FISCHER, K., BÖTTCHER, H. (1963): Die Sonnenkompaßorientierung der Eidechsen. Ergebn. Biol. **26**, 216–234.

BOVET, J. (1960): Experimentelle Untersuchungen über das Heimfindevermögen von Mäusen. Z. Tierpsychol. **17**, 728–755.

BOVET, J. (1968): Trails of deer mice *(Peromyscus maniculatus)* travelling on the snow while homing. J. Mamm. **49**, 713–725.

BOVET, J. (1971): Initial orientation of deer mice *(Peromyscus maniculatus)* released on snow in homing experiments. Z. Tierpsychol. **28**, 211–216.

BRAEMER, W. (1960): A critical review of the sun-azimuth hypothesis. Cold Spring Harb. Symp. **25**, 413–427.

BRAEMER, W., SCHWASSMANN, H. O. (1963): Vom Rhythmus der Sonnenorientierung bei Fischen am Äquator. Ergebn. Biol. **26**, 278–288.

BROWER, L. P. (1961): Studies on the migration of the monarch butterfly. I. Breeding populations of *Danaus plexippus* and *D. gilippus berenice* in south central Florida. Ecology **42**, 76–83.

BUCKLAND, F. (1880): Natural History of British Fishes. London: Unwin.

BULLOCK, T. H., RIDGWAY, S. H. (1972): Neurophysiological findings relevant to echolocation in marine animals. In: Animal Orientation and Navigation. S. R. GALLER, K. SCHMIDT-KOENIG, G. J. JACOBS, R. E. BELLEVILLE (eds.): National Aeronautics and Space Admin. Wash. D. C.: 373–395.

CARR, A. (1963): Orientation problem in the high-seas travel and terrestrial movements of marine turtles. In: Bio-Telemetry. L. SLATER (ed.): New York: Pergamon Press, 179–193.

CARR, A. (1972): The case for long-range chemoreceptive piloting in Chelonia. In: Animal Orientation and Navigation. S. R. GALLER, K. SCHMIDT-KOENIG, G. J. JACOBS, R. E. BELLEVILLE (eds.): National Aeronautics and Space Admin. Wash. D. C.: 469–483.

CLEAVER, F. C. (1964): Origins of high seas sockeye salmon. U. S. Fish. Bull. **63**, 445–476.

COCKRUM, E. L. (1956): Homing, movements and longevity of bats. J. Mamm. **37**, 48–57.

COOPER, R. A., UZMANN, J. R. (1971): Migrations and growth of deep-sea lobsters, *Homarus americanus*. Science **171**, 288–290.

CORTOPASSI, A. J., MEWALDT, L. R. (1965): The circumannual distribution of white-crowned sparrows. Bird Banding **36**, 141–169.

CRAIGHEAD, F. C. jr., CRAIGHEAD, J. J. (1965): Tracking grizzly bears. Biol. Science **15**, 88–92.

CRAIGHEAD, F. C. jr., CRAIGHEAD, J. J., COTE, C. E., BUECHNER, H. K. (1972): Satellite and ground radiotracking of elk. In: Animal Orientation and Navigation. S. R. GALLER, K. SCHMIDT-KOENIG, G. J. JACOBS, R. E. BELLEVILLE (eds.): National Aeronautics and Space Admin. Wash. D. C.: 99–111.

CRAIGHEAD, F. C. jr., CRAIGHEAD, J. J., DAVIES, R. S. (1963): Radiotracking of grizzly bears. In: Bio-Telemetry. L. SLATER (ed.): New York: Pergamon Press, 133–148.

CREUTZBERG, F. (1961): On the orientation of migrating elvers *(Anguilla vulgaris* Turt.) in a tidal area. Neth. J. Sea Res., **1**, 257–338.

CUMMINGS, W. C., PHILIPPI, L. A. (1970): Whale phonations in repetitive stanzas. Naval Undersea Res. and Dev. Center. San Diego, NVC TP 196.

CUMMINGS, W. C., THOMPSON, P. O., COOK, R. (1968): Underwater sounds of migrating gray whales, *Eschrichtius glaucus* (Cope). J. Acous. Soc. Am. **44**, 1278–1281.

DAWSON, C. E., IDYLL, C. P. (1951): Investigations on the Florida spiny lobster, *Panulirus argus* (Latreille) Marine Lab. Univ. Miami Tech. Ser. **2**, 1–39.

DEELDER, C. L., TESCH, F. W. (1970): Heimfindevermögen von Aalen *(Anguilla anguilla)*, die über große Entfernungen verpflanzt worden waren. Marine Biol. **6**, 81–92.

EDWARDS, R. Y., RITCEY, R. W. (1956): The migrations of a moose herd. J. Mamm. **37**, 486–494.

EGE, V. (1939): A revision of the genus Anguilla Shaw, a systematic, phylogenetic and geographical study. Dana Rep. 3, (16) 256–.

EMEIS, D. (1959): Untersuchungen zur Lichtkompaßorientierung des Wasserläufers *Velia currens* F. Z. Tierpsychol. **16**, 129–154.

EMLEN, S. T. (1972): The ontogenetic development of orientation capabilities. In: Animal Orientation and Navigation. S. R. GALLER, K. SCHMIDT-KOENIG, G. J. JACOBS, R. E. BELLEVILLE (eds.): National Aeronautics and Space Admin. Wash. D. C.: 191–210.

EMLEN, S. T. (1974): Migration: Orientation and navigation. In: Avian Biology. D. S. FARNER and J. R. KING (eds.): New York: Academic Press.

EMLEN, S. T., EMLEN, J. T. (1966): A technique for recording migratory orientation of captive birds. Auk **83**, 361–367.

ENRIGHT, J. T. (1961): Lunar orientation of *Orchestoidea corniculata* Stout (Amphipoda). Biol. Bull. **120**, 148–156.

ERCOLINI, A., SCAPINI, F. (1974): Sun compass and shore slope in the orientation of littoral amphipods (*Talitrus saltator* Montagu). Monit. Zool. Ital. (N. S.) **8**, 85–115.

FERGUSON, D. E. (1963): Orientation in three species of anuran amphibians. Ergebn. Biol. **26**, 128–134.

FERGUSON, D. E., LANDRETH, H. F. (1966): Celestial orientation of Fowler's toad *Bufo fowleri*. Behaviour **26**, 105–123.

FERGUSON, D. E., LANDRETH, H. F., TURNIPSEED, M. R. (1965): Astronomical orientation of the southern cricket frog, *Acris gryllus*. Copeia, 58–66.

FISCHER, K. (1961): Untersuchungen zur Sonnenkompaßorientierung und Laufaktivität von Smaragdeidechsen (*Lacerta viridis* Laur). Z. Tierpsychol. **18**, 450–470.

FISCHER, K. (1965): Sonnenkompaßorientierung und spontane Richtungstendenz bei jungen Suppenschildkröten (*Chelonia mydas* L.). Verh. Dt. Zool. Ges. **28**, 546–556.

FISLER, G. F. (1966): Homing in the western harvest mouse, *Reithrodontomys megalotis*. J. Mamm. **47**, 53–58.

FISLER, G. F. (1967): An experimental analysis of orientation to the home site in two rodent species. Canad. J. Zool. **45**, 261–268.

FORWARD, R. B. jr., HORCH, K. W., WATERMAN, T. H. (1972): Visual orientation at the water surface by the teleost zenarchopterus. Biol. Bull. **143**, 112–126.

FORWARD, R. B. jr., WATERMAN, T. H. (1973): Evidence for e-vector and light intensity pattern discrimination by the teleost Dermogenys. J. comp. Physiol. **87**, 189–202.

FRISCH, K. v. (1950): Die Sonne als Kompaß im Leben der Bienen. Experientia **6**, 210–221.

FRISCH, K. v. (1965): Tanzsprache und Orientierung der Bienen. Berlin—Heidelberg—New York: Springer.

FRISCH, K. v. (1968): Dance, Language and Orientation of bees. Cambridge: Harvard Univ. Press.

FROMME, H. G. (1961): Untersuchungen über das Orientierungsvermögen nächtlich ziehender Kleinvögel, *Erithacus rubecula*, *Sylvia communis*. Z. Tierpsychol. **18**, 205–220.

FUNKE, W. (1968): Heimfindevermögen und Ortstreue bei *Patella* L. (*Gastropoda, Prosobranchia*). Oecologia (Berl.) **2**, 19–142.

GIFFORD, C. E., GRIFFIN, D. R. (1960): Notes on homing and migratory behavior in bats. Ecology **41**, 378–381.

GOULD, E. (1957): Orientation in box turtles (*Terrapene c. carolinensis* Lin.). Biol. Bull. **112**, 336–148.

GRANT, D., ANDERSON, O., TWITTY, V. (1968): Homing orientation by olfaction in newts (*Taricha rivularis*). Science **160**, 1354–1356.

GRAUE, L. C. (1963): The effect of phase shifts in the day-night cycle on pigeon homing at distances of less than one mile. Ohio J. Sci. **63**, 214–217.

GREENWOOD, E. I., DURAND, J. A. (1955): The distribution of length and components of the sum of *n* random unit vectors. Ann. Math. Stat. *26*, 233–246.

GRIFFIN, D. R. (1952): Bird navigation. Biol. Rev. **27**, 359–400.

GRIFFIN, D. R. (1969): The physiology and geophysics of bird navigation. Quart. Rev. Biol. **44**, 255–276.

GROOT, C. (1965): On the orientation of young sockeye salmon (*Oncorhynchus nerka*) during their seaward migration out of lakes. Behaviour. Suppl. **14**, 1–198.

GRZIMEK, M., GRZIMEK, B. (1960): A study of the game of the Serengeti plains. Z. Säugetierkunde **25**, Sonderheft.

GWINNER, E. (1968a): Artspezifische Muster der Zugunruhe bei Laubsängern und ihre mögliche Bedeutung für die Beendigung des Zuges im Winterquartier. Z. Tierpsychol. **25**, 843–853.

GWINNER, E. (1968b): Circannuale Periodik als Grundlage des jahreszeitlichen Funktionswandels bei Zugvögeln. Untersuchungen am Fitis (*Phylloscopus trochilus*) und am Waldbaumsänger (*P. sibilatrix*). J. Ornithol. **109**, 70–95.

GWINNER, E. (1972): Endogenous timing factors in bird migration. In: Animal Orientation and Navigation. S. R. GALLER, K. SCHMIDT-KOENIG, G. J. JACOBS, R. E. BELLEVILLE (eds.): National Aeronautics and Space Admin. Wash. D. C.: 321–338.

GWINNER, E. (1974): Endogenous temporal control of migratory restlessness in warblers. Naturwiss. **61**, 405–406.

HARDEN JONES, F. R. (1965): Fish migration and water currents. Spec. Publ. Int. Comm. NW. Atlant. Fish **6**, 257–266.

HARDEN JONES, F. R. (1968): Fish Migration, London: Edward Arnold Ltd.

HARTT, A. C. (1962): Movement of the salmon in the North Pacific Ocean and Bering Sea as determined by tagging 1956–58. Int. North Pacific Fish. Comm. Bull. **6**, 157–.

91

PAPI, F., PARDI, L. (1953): Ricerche sull'orientamento di *Talitrus saltator* (Montagu) (Crustacea Amphipoda). II. Zeitschr. vergl. Physiol. **35**, 490–518.

PAPI, F., PARDI, L. (1959): Nuovi reperti sull'orientamento lunare die *Talitrus saltator* (Montagu) (Crustacea Amphipoda). Zeitschr. vergl. Physiol. **41**, 583–596.

PAPI, F., SYRJÄMÄKI, J. (1963): The sun-orientation rhythm of wolf spiders at different latitudes. Arch. ital. Biol. **101**, 59–77.

PARDI, L. (1957): Modificazione sperimentale della direzione di fuga negli anfipodi ad orientamento solare. Z. Tierpsychol. **14**, 261–275.

PARDI, L. (1960): Innate components in the solar orientation of littoral amphipods. Cold Spring Harb. Symp. **25**, 395–401.

PARDI, L., ERCOLINI, A. (1966): Ricerche sull'orientamento astronomico di Anfipodi litorali della zone equatoriale. III. L'orientamento solare in una populazione di *Talorchestia martensii* Weber a Sud dell' Equatore (4° Lat. S.) Monitore Zoologico Italiano **74**, 80–101.

PARDI, L., PAPI, F. (1952): Die Sonne als Kompaß bei *Talitrus saltator* (Montague) (Amphipoda Talitridae). Naturwiss. **39**, 262–263.

PARDI, L., PAPI, F. (1953): Ricerche sull' orientamento di *Talitrus saltator* (Montagu) (Crustacea Amphipoda). Z. vergl. Physiol. **35**, 459–489.

PERDECK, A. C. (1958): Two types of orientation in migrating starlings *Sturnus vulgaris* L., and chaffinches, *Fringilla coelebs* L., as revealed by displacement experiments. Ardea **46**, 1–37.

PRECHT, H., LINDENLAUB, E. (1954): Über das Heimfindevermögen von Säugetieren I. Versuche an Katzen. Z. Tierpsychol. **11**, 485–494.

RAINEY, R. C. (1951): Weather and the movements of locust swarms: a new hypothesis. Nature **168**, 1057–1060.

RAINEY, R. C. (1963): Meteorology and the migration of locusts. Tech. Notes Wld. met. Org. **54**, 115 pp.

RAO, J. S. (1969): Some contributions to the analysis of circular data. Ph. D. dissertation. Calcutta: Indian Statistical Institute.

RAWSON, K. S., RAWSON, A. M. (1955): The orientation of homing pigeons in relation to change in sun declination. J. Orn. **96**, 168–172.

REYSENBACH DE HAAN, F. W. (1966): Listening underwater: thoughts on sound and cetacean hearing. In: Whales Dolphins and Porpoises. K. S. NORRIS (ed.): Univ. of Calif. Press: 583–596.

RICKER, W. E. (1938): Residual and kokanee salmon in Cultus Lake. J. Fish. Res. Bd. Can. **4**, 192–218.

ROBINSON, W. L., FALLS, J. B. (1965): A study of homing of meadow mice. Amer. Midl. Nat. **73**, 188–224.

SANTSCHI, F. (1911): Observations et remarques critiques sur le mécanisme de l'orientation chez les fourmis. Rev. Suisse Zoologie **19**, 303–338.

SAUER, F. (1957): Die Sternenorientierung nächtlich ziehender Grasmücken *(Sylvia atricapilla, borin* und *curruca)*. Z. Tierpsychol. **14**, 29–70.

SAUER, E. G. F. (1961): Further studies on the stellar orientation of nocturnally migrating birds. Psychol. Forschung **26**, 224–244.

SAUER, E. G. F. (1963): Migration habits of golden plovers. Proc. XIII. Int. Ornithol. Congr. Ithaca, N. Y.: 454–467.

SAUER, F., SAUER, E. (1955): Zur Frage der nächtlichen Zugorientierung von Grasmücken. Rev. Suisse Zool. **62**, 250–259.

SAUER, E. G. F., SAUER, E. M. (1960): Star navigation of nocturnal migrating birds. The 1958 planetarium experiments. Cold Spring Harb. Symp. Quant. Biol. **25**, 463–473.

SAYER, H. J. (1956): A photographic method for the study of insect migration. Nature **177**, 226.

SAYER, H. J. (1965): The determination of flight performance of insects and birds and the associated wind structure of the atmosphere. Anim. Behaviour **13**, 337–341.

SCHLICHTE, H. J. (1973): Untersuchungen über die Bedeutung optischer Parameter für das Heimkehrverhalten der Brieftaube. Z. Tierpsychol. **32**, 257–280.

SCHLICHTE, H. J., SCHMIDT-KOENIG, K. (1971): Zum Heimfindevermögen der Brieftaube bei erschwerter optischer Wahrnehmung. Naturwiss. **58**, 329–330.

SCHMIDT, J. (1906): Contribution to the life-history of the eel (*Anguilla vulgaris* Flem.) Rapp. P.—v. Réun. Cons. perm. int. Explor. Mer. **5**, 137–274.

SCHMIDT, J. (1913): First report on eel investigation 1913. Rapp. P.—v. Réun. Cons. perm. int. Explor. Mer. **18**, 29 pp.

SCHMIDT, J. (1932): Danish eel investigations during 25 years 1905–1930. Copenhagen: Carlsberg Foundation.

SCHMIDT-KOENIG, K. (1958): Experimentelle Einflußnahme auf die 24-Stunden-Periodik bei Brieftauben und deren Auswirkungen unter besonderer Berücksichtigung des Heimfindevermögens. Z. Tierpsychol. **15**, 301–331.

SCHMIDT-KOENIG, K. (1960): Internal clocks and homing. Cold Spring Harb. Symp. **25**, 389–393.

SCHMIDT-KOENIG, K. (1961): Die Sonne als Kompaß im Heim-Orientierungssystem der Brieftauben. Z. Tierpsychol. **18**, 221–244.

SCHMIDT-KOENIG, K. (1963): Neuere Aspekte über die Orientierungsleistungen von Brieftauben. Ergebn. Biol. **26**, 286–297.

SCHMIDT-KOENIG, K. (1965): Current problems in bird orientation. In: Adv. Study Behaviour. D. LEHRMAN, E. HINDE, E. SHAW (eds.): 217–278. New York: Academic Press.

SCHMIDT-KOENIG, K. (1969): Weitere Versuche, durch Verstellen der inneren Uhr in den Heimkehrprozess der Brieftaube einzugreifen. Verh. Dtsch. Zool. Ges. **33**, 200–205.

SCHMIDT-KOENIG, K. (1970a): Ein Versuch, theoretisch mögliche Navigationsverfahren von Vögeln zu klassifizieren und relevante sinnesphysiologische Probleme zu umreißen. Verh. Dtsch. Zool. Ges. **64**, 243–245.

SCHMIDT-KOENIG, K. (1972): New experiments on the effect of clock shifts on homing in pigeons. In: Animal Orientation and Navigation. S. R. GALLER, K. SCHMIDT-KOENIG, G. J. JACOBS, R. E. BELLEVILLE (eds.): National Aeronautics and Space Admin. Wash. D. C.: 275–285.

SCHMIDT-KOENIG, K. (1973): Über die Navigation der Vögel. Naturwiss. **60**, 88–94.

SCHMIDT-KOENIG, K., MATTHEWS, G. V. T. (1970b): Bird Navigation, Cambridge Univ. Press 1968. Z. Tierpsychol. **27**, 118–120.

SCHMIDT-KOENIG, K., SCHLICHTE, H. J. (1972): Homing in pigeons with impaired vision. Proc. Nat. Acad. Sci. USA **69**, 2446–2447.

SCHMIDT-KOENIG, K., WALCOTT, CH. (1973): Flugwege und Verbleib von Brieftauben mit getrübten Haftschalen. Naturwiss. **60**, 108–109.

SCHMIDT-KOENIG, K., WALCOTT, CH. (1975): Tracks of pigeon homing with frosted lenses, in prep.

SCHÜZ, E. (1971): Grundriß der Vogelzugskunde. Berlin, Hamburg: Parey.

SIEBECK, O. (1960): Untersuchungen über die Vertikalwanderung planktischer Crustaceen unter Berücksichtigung der Strahlungsverhältnisse. Int. Revue ges. Hydrobiol. **45**, 381–454.

SIEBECK, O. (1968): "Uferflucht" und optische Orientierung pelagischer Crustaceen. Arch. Hydrobiol. Suppl. **35**, 1–118.

SIEBECK, O. (1973): Untersuchungen zur Biotopbindung einheimischer Pelagial-Crustaceen. Saarbrücken: Verh. Ges. Ökol. 11–24.

TAYLOR, D. H., FERGUSON, D. E. (1969): Solar cues and shoreline learning in the southern cricket frog, *Acris gryllus*. Herpetologica **25**, 147–149.

TAYLOR, D. H., FERGUSON, D. E. (1970): Extraoptic celestial orientation in the southern cricket frog *Acris gryllus*. Science **168**, 390–392.

TESCH, F. W. (1970): Heimfindevermögen von Aalen *Anguilla anguilla* nach Beeinträchtigung des Geruchssinnes, nach Adaptation oder nach Verpflanzung in ein Nachbar-Ästuar. Marine Biol. **6**, 148–157.

TESCH, F. W. (1974): Speed and direction of silver and yellow eels, *Anguilla anguilla*, released and tracked in the open North Sea. Wiss. Komm. Meeresforsch. **23**, 181–197.

TRACY, C. R., DOLE, J. W. (1969a): Orientation of displaced Californian toads, *Bufo boreas*, to their breeding sites. Copeia, 693–700.

TRACY, C. R., DOLE, J. W. (1969b): Evidence for celestral orientation in adult California toads, *Bufo boreas halophilus*. Bull. So. Calif. Acad. Sci. **68**, 10–18.

TUCKER, D. W. (1959): A new solution to the Atlantic eel problem. Nature **183**, 495–501.

TWITTY, V. (1959): Migration and speciation in newts. Science **130**, 1735–1743.

TWITTY, V., GRANT, D., ANDERSON, O. (1964): Long-distance homing in the newt *Taricha rivularis*. Proc. Nat. Acad. Sci. U. S. **51**, 51–58.

TWITTY, V. C. (1966): Of Sientists and Salamanders. San Francisco: Freeman.

URQUHART, F. A. (1958): A discussion of the use of the word 'migration' as it relates to a proposed classification for animal movements. Contr. R. Ontario Mus. Div. Zool. Paleont. No. 50, 11 pp.

URQUHART, F. A. (1960): The Monarch Butterfly. Toronto: University of Toronto Press.

WAGNER, H. (1968): Topographisch bedingte zweigipfelige und schiefe Kreisverteilungen bei der Anfangsorientierung verfrachteter Brieftauben. Revue Suisse de Zoologie **75**, 682–690.

WALCOTT, C. (1972): The navigation of homing pigeons: do they use sun navigation? In: Animal Orientation and Navigation. S. R. GALLER, K. SCHMIDT-KOENIG, G. J. JACOBS, R. E. BELLEVILLE (eds.): National Aeronautics and Space Admin. Wash. D. C.: 283–292.

WALCOTT, C., SCHMIDT-KOENIG, K. (1973): The effect on pigeon homing of anesthesia during displacement. Auk **90**, 281–286.

WALLRAFF, H. G. (1965): Über das Heimfindevermögen von Brieftauben mit durchtrennten Bogengängen. Z. vergl. Physiol. **50**, 313–330.

WALOFF, Z. (1946): A long-range migration of the desert locust from southern Marocco to Portugal, with an analysis of concurrent weather conditions. Proc. Roy. Ent. Soc. Lon. (A) **21**, 81–84.

WALOFF, Z. (1959): Notes on some aspects of the desert locust problem. In: Rep. of the FAO Panel of Aspects of the Strategy of Desert Locust Plague Control. FAO Document 59-6-4737: 23–26 cited after JOHNSON (1969).

WATERMAN, T. H. (1972): Visual direction finding by fishes. In: Animal Orientation and Navigation. S. R. GALLER, K. SCHMIDT-KOENIG, G. J. JACOBS, R. E. BELLEVILLE (eds.): National Aeronautics and Space Admin. Wash. D. C.: 437–456.

WATERMAN, T. H., FORWARD, R. B. jr. (1972): Field demonstration of polarotaxis in the fish Zenarchopterus. J. exp. Zool. **180**, 33–54.

WATKINS, W. A. (1966): Listening to cetaceans. In: Whales, Dolphins and Porpoises. K. S. NORRIS (ed.), Univ. of Calif. Press: 471–476.

WATSON, G. S. (1961): Goodness-of-fit test on a circle. Biometrika **48**, 109–114.

WATSON, G. S. (1962): Goodness-of-fit test on a circle II. Biometrika **49**, 57–63.

WATSON, G. S., WILLIAMS, E. J. (1956): On the construction of significance tests on the circle and the sphere. Biometrika **43**, 344–352.

WILLIAMS, T. C., WILLIAMS, J. M. (1967): Radio tracking of homing bats. Science **155**, 1435–1436.

WILLIAMS, T. C., WILLIAMS, J. H., GRIFFIN, D. R. (1966): The homing ability of the neotropical bat *Phyllostomus hastatus*, with evidence for visual orientation. Anim. Behavior **14**, 468–473.

WILTSCHKO, W. (1968): Über den Einfluß statischer Magnetfelder auf die Zugorientierung der Rotkehlchen *(Erithacus rubecula)*. Z. Tierpsychol. **25**, 537–558.

WILTSCHKO, W. (1973): Kompaßsysteme in der Orientierung von Zugvögeln. Akad. Wiss. Lit. Mainz, Reihe Inf. Org. II. Wiesbaden: Steiner.

WILTSCHKO, W., WILTSCHKO, R. (1972): The magnetic compass of European robins. Science **176**, 62–64.

WINN, H. E., SALMON, M., ROBERTS, N. (1964): Sun-compass orientation by parrot fishes. Z. Tierpsychol. **21**, 798–812.

YEAGLEY, H. L. (1947): A preliminary study of a physical basis of bird navigation. J. appl. Phys. **18**, 1035–1063.

YEAGLEY, H. L. (1951): A preliminary study of a physical basis of bird navigation II. J. appl. Phys. **22**, 746–760.

Subject Index

Zoophysiology and Ecology

Editors: D.S. Farner (Managing Editor), W.S. Hoar, J. Jacobs, H. Langer, M. Lindauer

Vol. 1: P.J. BENTLEY
Endocrines and Osmoregulation
A Comparative Account of the Regulation of Water and Salt in Vertebrates

Contents: Osmotic Problems of Vertebrates. – The Vertebrate Endocrine System. – The Mammals. – The Birds. – The Reptiles. – The Amphibia. – The Fishes.

Vol. 2: L. IRVING
Arctic Life of Birds and Mammals Including Man

Contents: Environment of Arctic Life. – Mammals of the Arctic. – Arctic Land Birds and Their Migrations. – Maintenance of Arctic Populations: Birds. – Maintenance of Arctic Populations of Mammals. – Warm Temperature of Birds and Mammals. – Maintenance of Warmth by Variable Insulation. – Metabolic Supply of Heat. – Heterothermic Operation of Homeotherms. – Size and Seasonal Change in Dimensions. – Insulation of Man.

Vol. 3: A.E. NEEDHAM
The Significance of Zoochromes

Contents: General. – The Nature and Distribution of Zoochromes. – Physiological Functions of Zoochromes. – Biochemical Functions of Zoochromes. – The Significance of Zoochromes for Reproduction and Development. – Evidence from Chromogenesis in the Individual. – Evolutionary Evidence and General Assessment.

Vol. 4/5: A.C. NEVILLE
Biology of the Arthropod Culticle

Contents: General Structure of Integument. – The Structural Macromolecules. – Molecular Cross-Linking. – Supermolecular Architecture. – Physiological Aspects. – Calcification. – Physical Properties. – Phylogenetical Aspects Aspects. – Outstanding Problems.

Springer-Verlag
Berlin
Heidelberg
New York

Related Titles

Effects of Temperature on Ectothermic Organisms
Ecological Implications and Mechanisms of Compensation
Edited by W. Wieser

This report is about the way ectothermic organisms react to temperature changes. The first part takes the biochemical-physiological view, discussing the molecular mechanisms of adaptation; the second part analyzes the manner in which an ecological factor — temperature — influences biological processes.

E. BÜNNING
The Physiological Clock
Circadian Rhythms and Biological Chronometry
(Heidelberg Science Library, Vol. 1)

The physiological clock has acquired particular significance with the advent of intercontinental air travel and space travel. This book gives a detailed discussion of present-day knowledge of the processes basic to this mechanism and the effects on the human organism.

The Early Life History of Fish
The Proceedings of an International Symposium Held at the Dunstaffnage Marine Research Laboratory of the Scottish Marine Biological Association at Oban, Scotland, from May 17-23, 1973
Edited by J.H.S. Blaxter

Studies on the early life history of fish have an important part to play in securing future food supplies. This volume represents a very full progress report on research being carried on on fish eggs and larvae, not only at sea but also increasingly in the laboratory.

Experimental Analysis of Insect Behaviour
Edited by L.B. Browne

This volume is a collection of papers representative of a great variety of approaches to the experimental analysis of insect behaviour. The approaches vary from the purely behavioural to the purely electrophysiological, from the identification of adequate stimuli to the origins, and from pure physiology to the experimental analysis of behavioural strategies.

Springer-Verlag
Berlin
Heidelberg
New York